Teshome Adugna, Girma Salale
Practical Chemistry

Also of interest

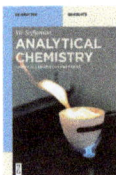

Analytical Chemistry.
Principles and Practice
Victor Angelo Soffiantini, 2021
ISBN 978-3-11-072119-5, e-ISBN 978-3-11-072120-1

Chemical Analysis in Cultural Heritage
Luigia Sabbatini and Inez Dorothé van der Werf (Eds.), 2020
ISBN 978-3-11-045641-7, e-ISBN 978-3-11-045753-7

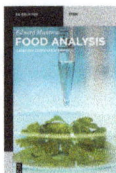

Food Analysis.
Using Ion Chromatography
Edward Muntean, 2022
ISBN 978-3-11-064438-8, e-ISBN 978-3-11-064440-1

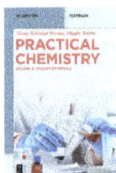

Practical Chemistry.
Transition Metals
Mesay Solomon Tesema, Digafie Zeleke, 2024
ISBN 978-3-11-157384-7, e-ISBN 978-3-11-157434-9

Teshome Adugna, Girma Salale

Practical Chemistry

Instrumental Analysis

Volume 1

DE GRUYTER

Authors
Teshome Adugna
Department of Chemistry
College of Natural Sciences
Salale University
P.O. Box 245
Oromia, Fiche
Ethiopia
teshyygoha@gmail.com

Dr. Girma Salale
Department of Chemistry
College of Natural Sciences
Salale University
P.O. Box 245
Oromia, Fiche
Ethiopia
ggirma245@gmail.com

ISBN 978-3-11-157504-9
e-ISBN (PDF) 978-3-11-157563-6
e-ISBN (EPUB) 978-3-11-157566-7

Library of Congress Control Number: 2024944100

Bibliographic information published by the Deutsche Nationalbibliothek
The Deutsche Nationalbibliothek lists this publication in the Deutsche Nationalbibliografie;
detailed bibliographic data are available on the internet at http://dnb.dnb.de.

www.degruyter.com
Questions about General Product Safety Regulation:
productsafety@degruyterbrill.com

———

To Derartu,
My love, my sweet, my life, my suninaf mam.
APRIL 2023
FICHE, ETHIOPIA

Contents

Laboratory manual

Back ground (course description)

Instrumental analysis laboratory I is an undergraduate course that covers the following instrumental methods of analysis: lab protocol orientation, chromatographic techniques (PC, paper chromatography; TLC, thin-layer chromatography; CC, column chromatography; GC, gas chromatography; and HPLC, high-performance liquid chromatography), and electroanalytical methods (potentiometry, voltammetry, conductometry, coulometry, electrogravimetry, electrophoresis, and refractive index).

Learning objectives/outcomes

With prior general knowledge of practical analytical chemistry, instrumental analysis (course prerequisites) is assumed for students of this level whom they are taking this laboratory.

    Upon successful completion of the course, students will be able to:

–   get hands-on experience or operate and calibrate different types of advanced instruments used for chemical analysis, including GC and HPLC;
–   identify different parts of selected instruments and describe their working principles;
–   compare, contrast, and discuss the applications of these instruments;
–   separate, identify, and determine the quantity of a given species from a sample by using different chromatographic techniques (PC, TLC, GC, HPLC, etc.); and
–   develop skills like being a team player through working in groups and technical writing skills through report writing with criteria that meet this stage.

Laboratory policy

It is required that students bring a lab coat, safety glasses, a notepad, and the lab manual with them when they arrive for the lab session. In addition to adhering to the university's standards for the delivery and assessment of courses, students are expected to actively engage in the learning process by doing as follows:

–   Respect all safety regulations in the lab
–   Dress appropriately
–   Maintain adequate hygiene, and turn in laboratory reports on time for each experiment
–   Save missing lab sessions for emergencies caused by illness or other legitimate issues

https://doi.org/10.1515/9783111575636-001

- Be prepared to learn and actively participate in laboratory works
- Save missing lab sessions for emergencies caused by illness or other legitimate issues
- Respect lab time; a maximum of 10-min delay for students to enter the lab is permitted
- Come prepared to learn and actively participate in laboratory tasks
- Bring a lab manual and flowchart

Grading/continuous evaluation/system:

Laboratory reports	30%
Evaluation/oral quizzes	10%
Practical and mid-term exam	20%
Final exams	40%

Laboratory reports

In addition to the answers to the experiment's questions and references, students will learn how to write a laboratory report in the official format, which consists of an introduction, an experimental section, results, a discussion, and a conclusion. Every student will get this competence by practicing the reports for every experiment that is conducted in the lab throughout the entire set of experiments.

Evaluation and quizzes

The laboratory technician and the lecturer will be responsible for conducting assessments of the students during the entire semester. Evaluation criteria include understanding of analytical chemistry, preparation and handling of solutions prior to instrumental measurement, pre-lab preparation, information gained from lab experiments, and ethics in the laboratory setting.

Oral examinations

This will test every student about his/her knowledge of the previous or past experiment he/she performed (theory, experimental part, laboratory report contents, and instrument component parts and their function). The quiz will be conducted through the laboratory session time, to measure the ability of students' understanding of basic ideas of the experiment. The questions of examinations will be based on the experiment theory, experimental part, and questions, and also what is discussed on the board during laboratory times. Laboratory reports contain all of this in addition to how to perform calculations related to the results of a certain experiment.

Laboratory experiments

The lab technician will divide the students into groups during the first week, with a minimum of five students per group. Every student in every group must turn in a laboratory report for the experiment they conducted, with each group performing one experiment during each lab period. Students receive a separate document with instructions on laboratory policy and the format for reports.

Laboratory notebook

A laboratory notebook is required to document your data and observations from the experiment. For this assignment, you can use a bound composition book. Attach your lab results with photocopies of the pertinent notebook pages.

Course outline and schedule

Week	Contents	Teaching methodology	Student activity
1st to 2nd	Lab protocol orientation	– Discussion – Demonstration – Rehearsal	– Attending – Participation – Manual collection
3rd	**Chromatography** – Determination of R_f of the given substance (dyes and amino acid) by paper and thin-layer chromatography using an organic solvent	– Brief lecture – Group discussion – Individual works – Experimentation – Demonstration	– Observational writing – Listening and observing – Demonstration and lecture
4th	**Chromatography** – Determination of R_f of a given dye (thymol blue, bromocresol, phenol red, etc.) using a solvent mixture	– Brief lecture – Group discussion – Individual works – Experimentation – Demonstration	– Take notes on the lesson treated – Ask questions on unclear idea – Active participation in discussion

(continued)

Week	Contents	Teaching methodology	Student activity
5th	– Determination of the number of constituents in a given mixture	– Brief lecture – Group discussion – Individual works – Experimentation – Demonstration	– Observational writing – Listening and observing – Demonstration and lecture – Take notes on the lesson treated – Ask questions on unclear idea – Active participation in discussion
6th	**Electrophoresis** – Determination of the charge and distance moved by an amino acid by the application of 300 V for a period of 1 h using an electrophoretic power supply	– Brief lecture – Group discussion – Individual works – Experimentation – Demonstration	– Observational writing – Listening and observing – Demonstration and lecture – Take notes on the lesson treated – Ask questions on unclear idea – Active participation in discussion
7th	– Determination of the number of amino acids in the given mixture by the electrophoresis method	– Brief lecture – Group discussion – Individual works – Experimentation – Demonstration	– Observational writing – Listening and observing – Demonstration and lecture – Take notes on the lesson treated – Ask questions on unclear idea – Active participation in discussion
8th	**Potentiometry** – Redox system: estimation of the given ferrous ammonium sulfate potentiometrically; a standard solution of 0.1 potassium dichromate solution may be provided	– Brief lecture – Group discussion – Individual works – Experimentation – Demonstration	– Observational writing – Listening and observing – Demonstration and lecture – Take notes on the lesson treated – Ask questions on unclear idea – Active participation in discussion

(continued)

Week	Contents	Teaching methodology	Student activity
9th	– Acid–base titration: estimation of hydrochloric acid potentiometrically using a calomel electrode	– Brief lecture – Group discussion – Individual works – Experimentation – Demonstration	– Observational writing – Listening and observing – Demonstration and lecture – Take notes on the lesson treated – Ask questions on unclear idea – Active participation in discussion
	– Determination of single electrode potential; silver, zinc, and copper electrodes may be used	– Brief lecture – Group discussion – Individual works – Experimentation – Demonstration	– Observational writing – Listening and observing – Demonstration and lecture – Take notes on the lesson treated – Ask questions on unclear idea – Active participation in discussion
10th	**Conductometry** – Acid–base titration: estimation of hydrochloric acid conductometrically using 0.5 N sodium hydroxide	– Brief lecture – Group discussion – Individual works – Experimentation – Demonstration	– Observational writing – Listening and observing – Demonstration and lecture – Take notes on the lesson treated – Ask questions on unclear idea – Active participation in discussion
11th	– Cell constant: determination of cell constant of a given conductivity cell using a conductivity meter – Equivalent conductance: determination of equivalent conductance of a given strong electrolyte	– Brief lecture – Group discussion – Individual works – Experimentation – Demonstration	– Observational writing – Listening and observing – Demonstration and lecture – Take notes on the lesson treated – Ask questions on unclear idea – Active participation in discussion
12th to 14th	Practical examination (20%)		
15th to 16th	Final examination (40%)		

Mode of assessment

Type	Description	%	Assessment date
Practical/skill evaluation	Individual/group performance and flowchart	10	Every week
Individual laboratory performance	Practical examination	20	Week 10
Experiment reports	Written report after each laboratory work	30	Every week
Written evaluation	Exam at the end of the laboratory work	40	At the end of the laboratory

Teacher's responsibilities

Teachers and teacher assistants should set a good example by donning personal protective equipment (PPE), adhering to and enforcing to safety policies and procedures, modeling safe behavior, and fostering a culture of safety. They ought to prioritize safety and take the initiative in all areas of laboratory safety. A checklist outlining crucial information for working in the high school laboratory is provided below for educators. This is a general safety checklist that needs to be updated and reviewed on a regular basis.

Upkeep of laboratory and equipment

As frequently as the management requests, conduct routine inspections of the safety and first aid supplies. On the connected equipment inspection tag, note the date of the inspection along with the initials of the inspector. If a potentially dangerous situation (such as a malfunctioning safety device or a chemical hazard) is found in the laboratory, report it in writing to the administration and keep them updated on the situation. Never operate faulty machinery, preserving records. For as long as the school system requires, maintain well-organized records of the staff's safety training. Follow procedures for safety and emergencies. Before beginning any laboratory activities, instruct students on where to find and how to utilize all safety and emergency equipment. Decide the safety measures to take in case of an accident or emergency.

Give pupils written and spoken safety instructions to refer to in case of an accident or emergency. Recognize where the water, gas, and electricity cutoff switches and valves are located and how to operate them in the laboratory. All safety and emergency equipment, such as fire blankets, eyewash, safety showers, fire extinguishers, and mercury spill kits, should be known where to find and how to use them. Keep a contact list for emergencies close at hand. Regularly carry out suitable safety and evacuation drills. Give children a thorough explanation of the repercussions for breaking safety regulations.

Maintenance of chemicals

Conduct routine chemical inventory inspections. Update the chemical inventory whenever the administration requests it, or at least once a year. Give the fire department and other local emergency responders a copy of the chemical inventory. Avoid storing food and beverages near chemicals. Preserve all chemicals in their original packaging whenever at all feasible. Verify the labels on all chemicals and reagents. Chemicals should not be kept in the laboratory chemical hood, on the floor, or on the lab bench. Make sure chemicals that are not in use are kept in a secured area with restricted access. Understand the needs for handling, storing, and disposing of each chemical that is utilized. Take care to dispose of chemicals appropriately. For information on disposal, refer to the label and the material safety data sheet, and always abide by the applicable chemical disposal laws.

Preparing for laboratory activities

Before beginning any laboratory exercise, balance the instructional value with any possible risks. Comprehend all possible risks associated with the materials, procedures, and tools used in each laboratory activity. Prior to usage, check all of the laboratory's apparatus and equipment. Give students instructions on all of the laboratory operations that will be carried out before they enter the room. Before beginning the assignment, go over any safety issues and possible risks associated with the laboratory work the students will be doing. Keep a record in your lesson plan book.

Ensuring appropriate laboratory conduct

Set an example of appropriate safety behavior that pupils can emulate. Make certain that students are donning the proper PPE, such as gloves, lab coats or aprons, and chemical splash goggles. Always follow the safety guidelines and protocols. Students should never be left unattended in the lab. The lab should never be opened to uninvited guests. Students should never be permitted to remove substances from the lab. Never allow gum, food, drinks, or smoking in the lab.

Evaluation of the course

Students will be reminded to provide the laboratory staff with anonymous written statements or comments at the end of the semester regarding the performance of the lab instructor, the technical assistance staff, the course material, the laboratory equipment, the facilities, etc. Your share of the courses benefits themselves.

1 Chromatography

1.1 Introduction

Chromatography, which comes from the Greek words "chroma," meaning "color," and "grafein," meaning "write," refers to a group of laboratory methods used to separate mixtures; that is, "color writing" is chromatography. One of the best analytical methods for separating or isolating organic compounds from a mixture is chromatography, which works by measuring the strength of the interaction between the stationary phase (SP; solid, liquid, or gas) and the mobile phase (solvent or gas), or MP. Chromatography is a useful technique for a variety of tasks including quantitative mixture analysis, kinetic studies of chemical reactions, pure substance synthesis, and molecular structure investigations on the molecular scale, quantitative determinations of mixtures, etc.

Sample mixes are forced through the SP's surface by the MP as it passes over the fixed SP. The SP is positioned fixedly and is used to divide mixtures into distinct components according to how the sample components interact with the MP. As the solutes have distinct affinities for the stationary and moving phases, separation happens when a mixture of solutes in a moving phase pass over a selectively absorbing matrix, the SP. Solutes with a higher affinity for the moving phase will stay in the moving phase longer than those with a higher affinity for the SP, which will cause them to travel more quickly. Classification of chromatographic methods are listed in table 1.

Thin-layer chromatography (TLC), column chromatography (CC), and paper chromatography (PC) are the three most popular chromatographic techniques. Proteins, enzymes, lipids, hormones, pigments, plant growth factors, and other naturally occurring organic compounds are routinely isolated using these procedures. Gas–liquid chromatography (GLC) and high-performance liquid chromatography (HPLC) are two types of more advanced chromatography that use instruments to separate molecules.

A dynamic and quick equilibrium between molecules of free (fully dissolved in the liquid or gaseous MP) and adsorbed (stuck to the surface of the solid SP) occurs in all kinds of chromatography. Chromatography provides quantitative information about individual solutes from peak height or area and retention times (R_ts) as well as qualitative information about spot area and retention factors (R_fs).

Calibration: To determine the amount of material (area under the peak/peak height) or relative proportions, known compounds are added to the column while the other conditions are maintained constant.

- Chromatographic methods can be broadly classified into four groups according to the nature of the two phases: They are
  - **Liquid chromatography:** It separates liquid samples with a liquid solvent (MP) and a column composed of solid beads (SP), e.g., HPLC, SEC, ion-exchange chromatography (IEC), IC, and CC.

https://doi.org/10.1515/9783111575636-002

Support material, SP, and solvent, or carrier gas, MP, are all that are required for chromatography.

Table 1: Classification of chromatographic methods based on MP and SP used for separation. Stock, R., Rice, C.B.F. (1974).

Chromatography type	Mobile phase	Stationary phase	Detectors	Applications	Examples
Paper chromatography	Vapor organic/ aqueous solvent	Cellulose paper	Optical detectors, UV–vis, fluorescence, etc.	Used for organic and colored compounds	PC
Thin-layer chromatography	Vapor organic/ aqueo us solvent	Silica gel supported on plastic film/ impregnated plates	Optical detectors, UV–vis, fluorescence etc.	Used for organic and colored compounds	TLC
Gas chromatography	Inert gases such as H_2, He, N_2, or Ar	Capillary or packed columns with substituted siloxanes	TCD, FID, MSD, fluorescence, etc.	Used for volatile organic and permanent gases	GLC and GSC
Liquid chromatography	Water and organic solvent	Capillary or packed columns with substituted siloxanes	UV–vis, RI, MSD, fluorescence, etc.	Used for thermally organic and inorganic compounds	HPLC, SEC, IEC, IC, and CC

- Based on the separation technique or principle, chromatography is classified as
  - Adsorption chromatography (LSC): Solutes adhered to are adsorbed onto SP surfaces through interactions between the SP's fixed active sites and the solute on a finely split solid adsorbent. Solute adsorption will be weaker, the higher the contact between the SP and the MP, and vice versa. For instance, GSC.
  - Partition chromatography (LLC): Solute species can equilibrate between two immiscible phases when dissolved in a liquid phase coated on a solid support surface. The solutes have a finite volume. A fixed (stationary) liquid phase is held in place by a finely split inert support (such as silica gel and kieselguhr) in the separating medium. Separation is accomplished by passing a MP over the SP. The SP can take the shape of a paper strip, a thin coating on glass, or a packed column. For instance, GLC and LLC.
  - Based on the relative polarities of the SP and MP, liquid–liquid chromatography can be divided into two phases:
    A) Normal-phase LLC: SP is polar and MP is nonpolar. Nonpolar solutes prefer the MP and elute first, while polar solutes prefer the SP and elute later.

B) Reversed-phase LLC: The SP is nonpolar, and the MP is polar, mostly water (H_2O). In such chromatography polar solutes prefer the MP and elute first, while nonpolar solutes prefer the SP and elute later.

The energy of interaction between a solute and the SP varies significantly depending on the functional groups. This variation gives rise to separations according to compound classes. A homogeneous series, on the other hand, is eluted as a single fraction from an adsorption column. Sometimes, the observed nonlinear relationship between sample concentration and the amount of sample being adsorbed can be avoided by deactivating the strongest adsorption sites (usually with water) and by keeping the amount of sample below the "linear capacity" of the adsorbent. Practically, compounds that have a high affinity for the MP emerge first (most volatile). The chromatograph's visual result is called a chromatogram. The retention time is displayed on the x-axis, while a signal representing the quantity of analyte leaving the system is plotted on the y-axis. Under some circumstances, the compound's retention time is distinctive. Column length, packing type, type of carrier gas, column temperature, and carrier gas flow rate are all factors that affect retention time.

– IEC involves cations that are covalently bonded to the SP and mobile anions. It is mostly the Coulombic attraction of a charged sample molecule by SP sites that are oppositely charged that serves as the retention mechanism. Either anion- or cation-exchanger chromatography can be used. The SP surface features charged functional groups such as $R–SO^{3-}$ and $N(CH_3)^+$. These species, which typically consist of porous, polymeric organic resins with different anionic or cationic exchanger groups, interact through electrostatic interactions.

– **Molecular (size) exclusion chromatography (SEC):** To segregate solutes according to their molecular size and shape, small molecules enter particle holes while large molecules are excluded. Large species cannot enter this chromatography, and small species may take longer to emerge from column packing. This chromatography functions similarly to a sieve phase. Gel permeation chromatography is another name for it.

  – **Affinity chromatography:** An immobilized species interacts with a specific solute in a very selective manner. This kind of chromatography technology is the most selective. To bring the mobile phase and the stationary phase into touch, there are two typical techniques. These are planar chromatography & column chromatography. a) planar chromatography: this type of chromatography uses a flat plate or paper as the support for the SP. Here, capillary action or gravity causes the MP to pass through the SP. It is made up of TLC, PC, and electrochromatography (EC).

  – b) column chromatography: **CC** is a kind of chromatography where the MP is pushed through the tube by gravity or pressure while the SP is contained in a small tube.

1.2 Paper chromatography (PC)

Theory

An analytical method for separating and identifying mixtures that are colored or can be colored, particularly pigments, is PC. Using an older method, a small dot of sample solution is applied to a strip of chromatography paper. The paper is placed into and sealed inside a jar with a thin layer of solvent. The sample mixture and the solvent meet when the solvent rises through the paper and begins to move up the paper together. In PC, a single solvent or a combination of solvents is employed as the MP, and filter paper made of pure cellulose serves as the SP. The sample mixture's various components interact with the paper over varying distances.

Partition chromatography is what PC is. The solute is split between the liquid MP and the liquid SP, which is made up of water adsorbed on the paper's hydrophilic surface. The technique is known as normal partition chromatography if the MP is less polar than the SP; in this scenario, polar compounds exhibit low R_f-values while nonpolar compounds exhibit high R_f values. The ratio of the distance traveled by the MP to the distance traveled by the sample component is known as the retention factor, or R_f:

$$R_f \text{ value} = \frac{\text{Distance traveled by substance}}{\text{Distance traveled by solvent front}}$$

The SP in reversed-phase partition chromatography is less polar than the MP. Polar chemicals have low R_f values, while nonpolar compounds have high R_f values due to this reason. One can achieve a reduction in the polarity of the SP by chemically treating the paper with sulfuric acid and acetic acid anhydride to generate acetyl cellulose, or by impregnating it with silicon or paraffin oil. R_f is used to identify mixtures of unknown substances. $0 < R_f \leq 1$; solutes cannot travel above or beyond the MP because they are carried by it.

Temperature, pH, solvent front/distance, type and quality of paper and water used as the SP, direction of fibers (horizontal, vertical, diagonal, zigzag, etc.), chromatographic development, solute spot, drying method, chemical reactions, and concentration of the separated substances are just a few of the variables that can affect R_f. TLC is frequently chosen because PC is less sensitive, takes longer to achieve a decent separation, and produces fewer precise spots overall. Separations of various dyes are shown in figure 1.

Either ascending chromatography, in which the solvent is filled at the bottom of the developing chamber and moves in an upward flow with gravity and capillary forces in the opposite direction, or descending chromatography, in which the MP moves downwards with gravity and capillary forces in the same direction while the paper is hanging in the chamber and touches with its upper edge in the solvent that is filled in a trough, can be used to develop a paper chromatogram. Due to the opposing effects of capillary action and gravity, the MP flows more slowly upward than downward.

It can be used for a variety of qualitative, semiquantitative, and preparative tasks, including figuring out what pigments are in plants, finding pesticides or insecticides in

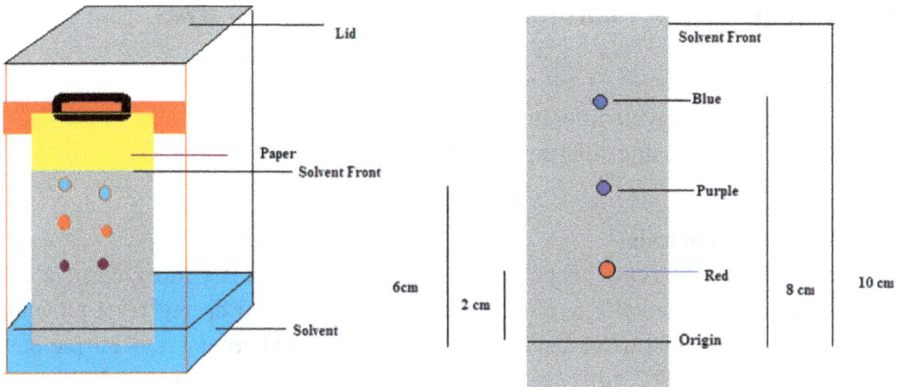

Figure 1: Diagram of paper chromatography. Byjus et al., 2009.

food, forensics dye composition analysis, identifying the compounds in a given substance, and sorting and testing complex mixtures of similar compounds, such as dyes, amino acids, anions, RNA, fingerprints, histamines and antibiotics. PC is a technique for dissecting mixtures into their constituent components. Pigments, dyes, and inks have historically been separated using PC. Paper with a lot of cellulose fiber, a fiber found in wood, and chromatography solution, which is often produced from a mixture of water and alcohol are required for PC.

Experiment 1: paper chromatography of indicator dyes

Objective
–   To get acquainted with different techniques of PC.
–   To separate dyes using different MPs.

Apparatus and chemicals
Developing chambers, ruler, pencil, measuring cylinders, capillary tubes, chromatography paper, dyes (methyl red, methyl violet, Congo red, and bromocresol purple), bromocresol green, mixture of methyl red, methyl violet, and bromocresol purple, solvent [MP 1: acetone, MP 2: ethyl acetate, and MP 3: acetone:ethyl acetate (50:50)].

Procedure
–   Prepare three tanks/chambers and fill them with different solvents.
–   Pour 50 mL of the solvent into the bottom of the tank and replace the lid.
–   The tank must always be covered to avoid the evaporation of solvent and to allow the atmosphere to become saturated.
–   Cut a sheet of paper in the format of 12 cm × 14 cm into strips for applying five spots, each 1.6 cm apart.
–   Draw a horizontal pencil line, about 2.0 cm from the strip's edge, along one of the shorter sides. This will serve as your "baseline," the initial point at which samples are detected and then utilized in the R_f value computations. The chromatography paper won't be carried up with graphite.
–   The first is 2.0 cm from the side and takes up a space of 1.6 cm between two successive spots. Apply the test solutions on the dot or five spots point using capillary tubes or micro pipettes. (NB: The spot should not be greater than 7 mm in diameter to prevent mixing of spots or strokes.)
–   Properly insert the paper without moving it into solvent holding tanks (don't shake the solvent) and close them.
–   Wait until the solvent front is 2 cm below the upper edge, i.e., the solvent reaches the top line edge.
–   After that, take the cellulose/sheet out of the tanks, dry the lower edge with soft tissue or filter paper to remove any leftover solvent, and expose the cellulose to the sun or air, if needed.
–   Draw a pencil line along the solvent front, then circle the sample sites that are at varying distances from the bottom front edge, which is 2 cm.
–   Measure the distance between the center of the circled sample spot and the lower front 2 cm edge. Finally, calculate the R_f value of each dye component and describe its characteristics.

Observation and calculation of data

Sample type (dyes)	Distance moved by MP	Distance moved by MP	R_f values
A			
B			
C			
D			
Unknown			

Questions
1. Why not use ink instead of a pencil for lining the baseline?
2. What is the importance of mixing different solvents as a MP?
3. What is the difference in distance moved by sample components in a single MP compared to a mixture of different miscible MPs?
4. What kind of cellulose is employed as the SP, and what kind of MP is it?
5. What are the limitations of PC's limitations?

1.3 Thin-layer chromatography (TLC)

Theory
TLC is a solid–liquid method of chromatography. In this method, the MP can be a single solvent or a mixture of solvents, depending on the material used in the SP. Typically, the SP consists of finely ground alumina ($Al_2O_3 \cdot xH_2O$) or cellulose immobilized onto a flat, plastic, inert carrier sheet made of glass or silica gel ($SiO_2 \cdot xH_2O)_n$ particles. The solute's distribution between the mobile and SPs, or its adsorption on the SP, serves as the basis for the separation. A thin layer of the specific SP is created by coating this absorbent on a glass slide or plastic sheet. The thin-layer plate can be heated in an oven to initiate the SP. A highly polar SP is produced by the electronegative oxygen and the electropositive silicon or aluminum. As a result, the SP will be more strongly attracted to the polar molecules that need to be separated. Almost any solvent mixture can serve as the MP. The MP can be changed to separate organic molecules. Similar to PCs, TLCs' R_f value is determined by the same principles and factors. If the components of the sample are colorless and difficult to detect, the spots can be identified using UV light or by reacting, spraying, or combining substances that generate colored compounds (such as ninhydrin solution) with the solutes under investigation.

It is possible to achieve good circumstances for separation when the R_f values fall between 0.2 and 0.8. The ratio of sample distance traveled to standard compound distance traveled on the same chromatogram is known as standard resolution or R_s. The values of R_s may be less than, greater than, or equal to 1. Only chromatographic studies conducted under the same conditions – such as temperature, SP, MP, developing chamber size, and development technique – can yield equivalent R_f and R_s results.

R_s is the ratio of the sample's journey time to the standard compound's travel time.

Generally speaking, molecules will move more slowly and have a stronger attachment to the SP when the functional group is more polar. Under severe conditions, molecules exhibit complete immobility. The solution to this issue is to make the MP more polar, which will alter the balance between the free and absorbed states in favor of the free.

Functional group-based elution sequence (using CC or silica or alumina TLC): fastest/highest (eluted using nonpolar mobile phase)

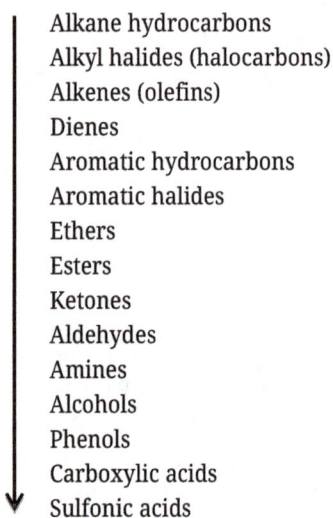

Alkane hydrocarbons
Alkyl halides (halocarbons)
Alkenes (olefins)
Dienes
Aromatic hydrocarbons
Aromatic halides
Ethers
Esters
Ketones
Aldehydes
Amines
Alcohols
Phenols
Carboxylic acids
Sulfonic acids

Lowest/slowest (needs polar mobile phase to elute)

TLC is a rapid, low-cost, microscale method that can be applied to
- Count the components in a mixture;
- Confirm the identity of a substance;
- Track the development of a reaction;
- Choose suitable CC conditions;
- Examine the fractions produced by CC;
- Identify the pigments present in a plant;
- Find pesticides or insecticides in food is a quick, inexpensive microscale technique that can be used to examine fiber dye composition in forensics;
- Identify the chemicals that are present in a specific sample;
- Quickly determine an organic compound's purity.

Using capillary action, the solution to be separated is dissolved in a suitable solvent and dragged up the plate. The experimental solution is then separated according to the size, polarity, solubility, and adsorption of the compound's constituent parts.

Experiment 2: thin-layer chromatography of dyes on silica gel

Objectives
- To become acquainted with different techniques of PC.
- To separate dyes using different MPs.
- To see the influence of solvent polarity and pH on the chromatographic separation.

Apparatus and chemicals
Developing chambers, precoated TLC plates, capillary tubes, acid dyes (bromocresol purple, bromophenol blue, methyl orange), basic dyes (methyl violet, malachite Green, methylene blue), mixture of dyes, solvents (MP 1: *n*-butanol: acetic acid: water in a 50:10:20 ratio and MP 2: *n*-butyl acetate: pyridine: water in a 30:45:25 ratio).

Procedure
- Prepare the two MPs by mixing the solvents in the given volume ratio using graduated cylinders.
- Fill the developing chambers to 0.5 cm high and close the lid for vapor saturation. In the meantime, apply the samples on the TLC plates.
- Place the samples as tiny spots on each plate where you have drawn a pencil line 2 cm above the plate's edge.
- Place the plates in the tanks and close them.
- Hold off until the solvent front is roughly 2 cm below the top border.
- Next, take the plates out of the tanks, use filter paper to remove any extra solvent from the lower edge, and use a pencil to label the solvent front.
- After the plates are dry, mark the spots on them and calculate the distance from the origin to the center of each spot.
- Then calculate the R_f values of all dyes in both MPs and determine the components of the mixtures.
- Finally, discuss the results with respect to the two MPs used.

Observation and calculation of data

Sample type (dyes)	Distance moved by MP	Distance moved by MP	R_f values
A			
B			
C			
D			
Unknown			

Questions

1. What are the properties of silica gel $(SiO_2 \cdot xH_2O)_n$ layers?
2. What detection methods exist for spots that are colorless?
3. Why is it necessary to close the tanks during the development of the chromatograms?
4. How does TLC differ from PC in terms of advantages? Briefly discuss.
5. How does the separation of basic or acidic dyes affect TLC?

Experiment 3: thin-layer chromatography of amino acid separations

Objective
- To separate amino acids in two different MPs.
- To see the influence of the solvent polarity on the R_f values of the amino acids.

Apparatus and chemicals
Developing chambers, precoated TLC plates, capillary tubes, atomizer, amino acids (leucine, lysine, aspartic acid, and alanine) and mixtures of amino acids, locating reagent: ninhydrin in acetone, solvents (MP 1: ethanol: water in a 70:30 ratio), and (MP 2: n-butanol: acetic acid glacial: water in 80:20:20 ratio).

Procedure
- Prepare the two MPs by mixing the solvents in the given volume ratio using graduated cylinders.
- Fill the developing chambers to 0.5 cm high and close the lid for vapor saturation. In the meantime, apply the samples on the TLC plates of dyes.
- On each plate, make a pencil line 2 cm above the edge. Then, apply the samples as tiny spots.
- After placing the plates within the tanks, close them.
- Once the solvent front is approximately 2 cm below the upper edge,
- then take the plates out of the tanks, use filter paper to remove any extra solvent, and mark the solvent front with a pencil. Measure the distance between the center of the spot and the origin.
- then calculate the R_f values of all dyes in both MPs and determine the components of the mixtures.
- Finally, discuss the results with respect to the two MPs used.

Observation and calculation of data

Sample type (dyes)	Distance moved by MP	Distance moved by MP	R_f values
A			
B			
C			
D			
Unknown			

1.4 Gas–liquid chromatography (GLC)

Theory
A liquid SP and a gaseous MP – usually an inert gas like helium or a nonreactive gas like nitrogen – are used in GLC, a separation technique. The basic idea behind separation is based on the variation in partition coefficients that volatilized molecules display between the liquid and gaseous phases. When a carrier gas carries the desired component through the column, this phenomenon is observed. A common use of GLC is the assessment of a substance's purity. Other common uses include the identification and separation of a mixture's constituent parts as well as the preparative chromatography technique, which creates novel molecules from mixtures.

The separation is based on the distribution of the sample between the stationary liquid phase and the MP. This can be expressed by the partition coefficient, K':

$$K' = [X]_s / [X]_m$$

where $[X]_s$ is the concentration of the sample in the SP, and $[X]_m$ is the concentration of the sample in the MP.

The partition coefficient depends only on the nature of the two phases and the sample. At low concentrations, like GC, the portion coefficient is often independent of the sample concentration. One then speaks about "linear" chromatography since the distribution isotherm is a straight-line function.

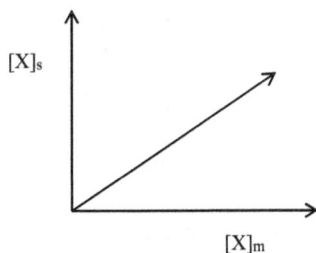

For the ideal case of any chromatographic separation, one assumes that
1. The instant equilibrium is maintained throughout the column at all times.
2. The transport of the sample is only by the motion of the carrier gas.
3. The column is packed uniformly.

There are different theories that provide a qualitative description of chromatographic separation. Two of them should be described here in brief detail: the plate theory and the rate theory.

Every chromatographic system consists of a SP that is coated onto the column or on the resin beads in the column, and a MP that moves the analytes through the column. Analytes are separated as a function of time spent in the separation column as a result of the SP's loose interactions with each analyte based on its chemical structure. Analytes move through the system more quickly, the less they interact with the SP. For less mobile analytes with stronger interactions, the opposite is true. As a result, under specific system conditions, the retention duration of a sample allows for the identification of its numerous analytes. These parameters in GC include temperature, linear velocity, flow rate, and gas (MP) pressure.

Plate theory

It is hypothesized that the column is divided into several zones known as theoretical plates. Assuming that the gas and liquid phases within each plate are in perfect equilibrium, the zone thickness or height equivalent to a theoretical plate (HETP) is calculated. The following presumptions are used to calculate the plate column's ensuing behavior:

1. The presence of additional solutes has no effect on the partition coefficient.
2. The distribution isotherm is linear.
3. Diffusion of the solute in the MP from one plate to another is neglected.

A model represents a chromatographic column as a sequence of distinct, thin portions known as theoretical plates. Assume that the analyte's equilibrium between the mobile and SPs is established at each plate. Analyte and MP movement is thought of as a succession of transfers from one plate to the next. A column's efficiency rises with the number of equilibrations, or theoretical plates, in the column.

A column is mathematically identical to a distillation plate column, according to plate theory. Each segment of the entire length represents a single equilibrium stage or theoretical plate. Every step establishes equilibrium. The number of theoretical plates in a column can be ascertained when a peak is created. The number of plates can be calculated using the peak's width and retention duration, assuming a Gaussian distribution.

As a result of the plate theory, the following equations are obtained, which give a qualitative description of theoretical plates is given by

$$N = 16 \, (tR/\Delta t)2$$

where tR is the retention time for a certain component and Δt is the base width of the peak of this compound.

A straightforward experiment can be used to calculate the theoretical plate count. A tiny volume of material is injected into the column, and measurements are made of the base width of the peak and retention time after the chromatogram is recorded. For each chemical, there is a separate theoretical plate count. A high separating power is shown by a large number, indicating that the column is well-suited for the separation of this compound.

When the column length is known, the HETP can be computed using the equation:

$$\text{HETP}: L/N = L/16\,(tR/\Delta t)2$$

where L is the length of column.

The smaller HETP indicates better efficiency of the column.

Rate theory

Effects affecting column performance such as
- phase characteristics,
- phase thickness,
- solute diffusivities,
- support size,
- support porosity,
- partition coefficients,
- phase velocity, and
- flow rates, can be predicted using rate theory.

Focuses on the roles that different kinetic factors play in band or zone broadening. It is believed that the column dispersivity, H, is the total of the kinetic components' separate contributions. The equation was written as follows by Van Deemter:

$$H = A + B/u + Cu \,(\text{van Deemter equation})$$

The reliance of the HETP on parameters such as the flow rate, the hydrodynamics of the MP, the rate of mass transfer between the stationary and MPs, and the solute diffusion rate along the column is the subject of rate theory. Zone spreading is caused by three phenomena: local nonequilibrium, eddy diffusion, and ordinary diffusion.

Normal diffusion: Solute molecules travel axially in the column from higher concentration regions to lower concentration regions. Diffuseness of zones is directly caused by this diffusion.

Eddy diffusion: Distinct, lengthy routes are taken by solute molecules as they pass through a porous material. While some molecules follow longer routes and fall behind the average, others follow shorter ones and advance the average. Additionally, the mobile path's velocity varies depending on where in the porous material it is located.

Local nonequilibrium: A solute's concentration profile in a specific zone resembles a bell-shaped Gaussian curve. This zone migrates throughout the column, bringing a constantly fluctuating concentration of solute in touch with a particular area of the column. The solute's concentration in the MP varies with time, making it unable to properly establish the distribution and/or adsorption desorption equilibrium.

The van Deemter equation in chromatography, named for Jan van Deemter, relates the variance per unit length of a separation column to the linear MP velocity by considering the physical, kinetic, and thermodynamic properties of a separation. These properties include pathways within the column, diffusion (axial and longitudinal), and mass transfer kinetics between stationary and MPs. In liquid chromatography, the MP velocity is taken as the exit velocity, that is, the ratio of the flow rate in milliliter/second to the cross-sectional area of the "column-exit flow path." For a packed column, the cross-sectional area of the column exit flow path is usually taken as 0.6 times the cross-sectional area of the column. Alternatively, the linear velocity can be taken as the ratio of the column length to the dead time. If the MP is a gas, then the pressure correction must be applied. The ratio of the column length to the column efficiency in theoretical plates is used to calculate the variance per unit length of the column. According to the hyperbolic function known as the van Deemter equation, there exists an optimal velocity that yields the lowest variance per unit column length and, hence, the highest efficiency. When the first time rate theory was applied to the chromatography elution process, the van Deemter equation was produced.

Highest point expanding results from multiple flow and kinetic parameters, which are related to the HETP of a chromatographic column using the van Deemter equation as follows:

$$\text{HETP} = A + \frac{B}{u} + (C_s + C_m) \times u$$

where HETP is a measurement of the column's resolving power (m), A the Eddy-diffusion parameter [m] in relation to channeling through a nonideal packing, B is the eluting particles' longitudinal diffusion coefficient, which produces dispersion [m^2 s^{-1}], and C is the analyte's resistance to the mass transfer coefficient between the stationary and MPs [s]: u = speed [m/s].

Figure 2 illustrates how the HETP depends on the flow rate. It is evident that the HETP approaches a minimum and maximum efficiency (HETPmin) at an ideal flow rate (u_{opt}). The column's strongest separating power is at u_{opt}.

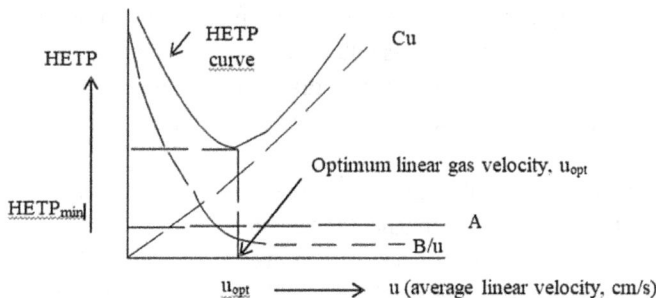

Figure 2: Dependence of HETP on flow rate according to the van Deemter equation.

Instrumentation

A schematic diagram of gas chromatography is shown in figure 3.

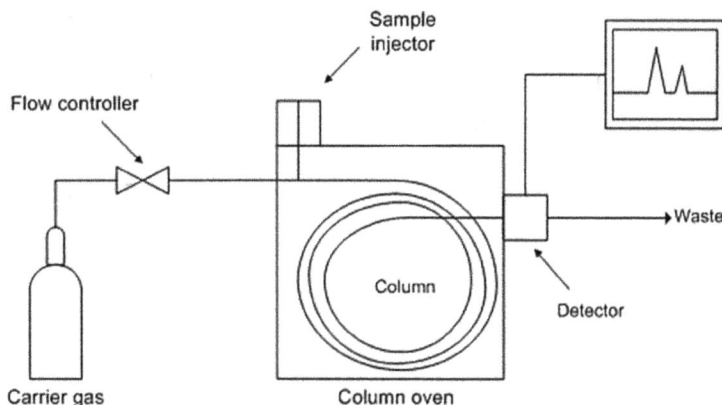

Figure 3: Schematic diagram of the components of a typical gas chromatograph. Adapted from http://en. Wikipedia.org/wiki/Gas_chromatography.

The huge gas cylinder provides the inert carrier gas (such as He or N2). Pressure regulators are used in both the GC and the gas cylinder to regulate the MP's flow rate. The sample is injected into the sample injection port using a syringe, where it is rapidly vaporized and transported into the column. In order to prevent band spreading and inadequate separation, a rapid sample introduction is required. By maintaining the injection temperature above the boiling point of the sample's constituent parts, a quick vaporization of the sample is achieved.

In order to regulate the column's temperature, it is placed inside an oven. In GC, two kinds of columns are employed: capillary and packed columns.

A glass, metal, or Teflon tube with a diameter of 2–6 mm and a length of 2–4 m makes up a packed column. In order to fit into a thermostat – which is used to change the column temperature if the sample needs to be analyzed at a temperature other than room temperature – the tubes are typically U-shaped or coiled. The tube is filled with the solid support covered in the stationary liquid phase.

Capillary tubing is used to create a capillary column. The liquid phase is deposited inside the tubing in extremely thin layers. A capillary column's pressure loss is quite minimal. As a result, long columns with many potential plates can be created.

At room temperature, the flow at the detector's outlet is used to calculate the carrier gas's flow rate. In GLC, flow meters known as soap film flow meters are typically used.

Detector

A chromatogram is created when an electrical signal is amplified and transformed from the detector's sense of a physicochemical feature of the analyte. With the exception of the mass spectrometer and the thermal conductivity detector (TCD), the majority of detectors used in GC were created expressly for this method. About sixty detectors have been employed in GC overall. "Selective detectors" are those that react more strongly to specific types of analytes. The application of GC in conjunction with mass spectrometry (MS) has grown over the past 10 years. The mass spectrometer has evolved into a common detector that does not require the separation of every component in the sample and permits lower detection limits. One is mass spectroscopy.

In GC, the detector needs to show that there is a trace amount of eluted sample present in a large amount of carrier gas. As a result, the detector needs to be quick to react and sensitive to a range of chemicals. When considering the concentration of the various sample components, the detector signal ought to follow a linear pattern.

Many different detectors are used in GC. The most common are the thermal conductivity and the flame ionization detectors (FIDs), which are shown below in table 2.

Table 2: Different types of detectors in gas chromatography.

Name of detector	Type	Selectivity	Minimum detectability	Linear range
Flame ionization (FID)	Universal	A	10 pg C/s	10^7
Thermal conductivity (TCD)	Universal	B	1 ng/mL mobile phase	10^5
Electron capture (ECD)	Selective	Compounds capturing electrons, e.g., halogens	0.2 pg Cl/s	10^3
Nitrogen phosphorus (NDP)	Selective	N and P	1 pg N/s, 0.5 pg P/s	10^4
Flame photometric (FPD)	Selective	P and S	50 pg S/s 2 pg P/s	10^3 10^4
Photoionization (PID)	Selective	Aromatics	5 pg P/s	10^7
Electrolyte conductivity (ELCD)	Selective	Halogens and S	1 pg Cl/s 5 pg S/s	10^6 10^4
Fourier transform infrared spectroscopy (FTIR)	Universal	Molecular vibrations	1 ng (strong absorber)	10^5
Mass spectroscopy (MSD)	Universal	Characteristic ions	1 ng full-scan mode 1 pg ion-monitoring mode	10^5
Atomic emission (AED)	Universal	Any element	0.2–50 pg/s depending on element	10^4

Thermal conductivity detector

This kind of universal detector has 10^6 linear ranges and a MP of 1 ng/mL. This ubiquitous, nondestructive, concentration-sensitive detector is the most significant of the bulk physical property detectors. When the sample goes through the detector, it detects a change from the carrier gas's baseline response. A heated metal filament is inserted into a metal block cavity with an appropriate heat capacity to form the detector. The gas has a major influence on the filament's temperature. Variations in the gas's thermal conductivity owing to the presence of sample components cause the filament's temperature to fluctuate, which in turn causes variations in the hot filament's resistance. Hydrogen and helium are the best carrier gases to use in conjunction with this type of detector since their thermal conductivities are much higher than any other gases; on safety grounds, helium is preferred because of its inertness.

Two matching pairs of filaments are arranged in a Wheatstone bridge circuit in the detector; the chromatographic column's effluent surrounds the other two filaments, while the carrier gas only surrounds the filaments in the opposite arms of the bridge. There were two gas channels installed in the cell: a reference channel and a sample channel. The bridge is balanced when pure carrier gas travels over the reference and sample filaments; however, the bridge becomes imbalanced when vapor appears from the column, changing the rate at which the sample filaments cool. The amount of vapor in the carrier gas at that particular moment and the out-of-balance signal over the bridge determine how much of this imbalance there is. This is a measure of the concentration of vapor in the carrier gas at that instant, and the out-of-balance signal across the bridge is measured and displayed on a recorder in the form of chromatogram.

The differential technique used is thus based on the measurement of the difference in thermal conductivity between the carrier gas and the carrier gas sample mixture. TCD is generally used for the detection of permanent gases, light hydrocarbons, and compounds that respond poorly to the FID (see figure 4).

The TCD has been designed to be used with capillary columns, with a small volume to meet the requirement that the detector's response time be much smaller than the chromatographic peak width. The practical problem that TCDs become increasingly sensitive to external impacts when cell capacity is decreased competes with this aim for compact volume.

Flame ionization detector

In order to create a flame, hydrogen is combined with the carrier gas that exits the column and holds the organic vapors. Low ionization potential solute molecules become ionized in the flame. The generated ions are gathered at electrodes, and the ion current that results is then measured. The negative electrode is formed by the burner jet. The positive electrode is a grid of wires that extends into the flame's tip. The most popular GC detector is the FID. Its operation is straightforward, response time is quick, sensitivity is on the order of 10 pg carbon per second with a linear range of 10^7,

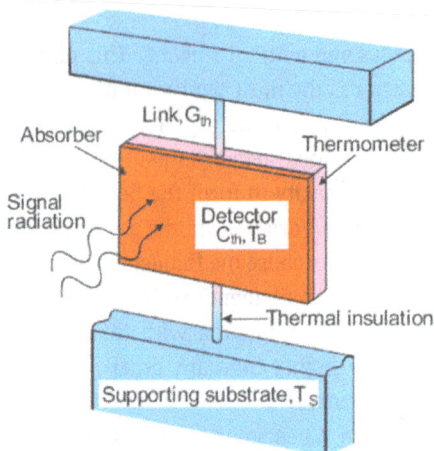

Figure 4: Schematic circuit of a thermal conductivity detector. John V. Hinshaw, 2006.

and detector stability is outstanding. Hydrogen is combined with the carrier gas, and this combination is combusted in air at the exit of a flame jet.

The fundamental mechanism of this detector involves combining column effluent with hydrogen and burning it in air to create a flame strong enough to ionize solute molecules with low ionization potentials. The burner jet is the negative electrode, while the anode is often a wire or grid extending into the flame tip. The ions created are collected at electrodes, and the resulting ion current is measured. Very few ions are produced during the combustion of hydrogen and air mixtures, resulting in an almost steady signal when only the carrier gas and hydrogen burn.

However, ionization and a significant rise in the flame's electrical conductivity occur when carbon-containing substances are present. When additional analysis of the eluate is required, stream-splitting devices are used since the sample is destroyed in the flame. This device is placed between the column and detector and permits the majority of the sample to bypass the detector (see figure 5).

With its practically universal use for GC of organic substances, the FID's high sensitivity, stability, quick reaction, and broad linear response range (~10^7) have made it the most widely used detector available today.

Advantages of flame ionization detector
– A similar sensitivity to the majority of organic substances,
– No reaction to typical contaminants in carrier gases,
– Little impact from variations in pressure, temperature, or flow,
– Consistent baseline,
– Cost: Flame ionization detectors can be purchased and used for comparatively little money.
– Minimal maintenance needs: These detectors require very little care, aside from the occasional cleaning or replacement of the FID jet.

– Robust design: FIDs are comparatively impervious to abuse

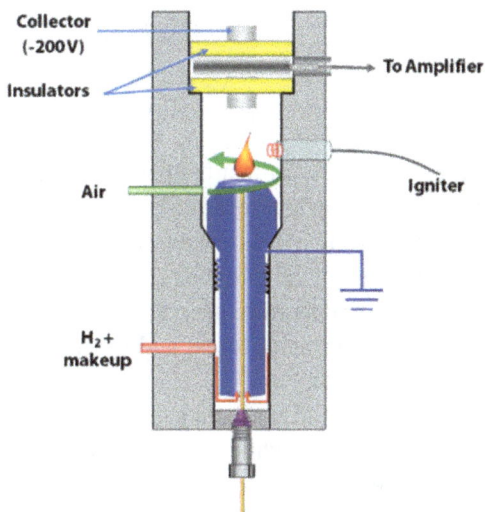

Figure 5: Flame ionization detector.

There are numerous applications for GC, including the detection of bombs in airports and the comparison of victim fibers in forensics. It is employed to examine body fibers and blood samples recovered from crime scenes. In addition, it is utilized for the analysis of volatile gases and the identification of elements in organic and organometallic samples including carbon, hydrogen, nitrogen, oxygen, and sulfur.

Helium (He) is utilized in GC to force the gaseous mixture through an absorbent material column. These methods can be used to quantify peak area, which is frequently employed as a quantitative indicator of a certain component in the sample.

GC sample preparation
The standard GC sample for somewhat volatile compounds can be prepared using two methods based on the types of samples (liquid and solid). These are:

Liquid sample preparation
1) Dip a Pasteur pipette's tip into the liquid. Approximately 10 mg of the liquid will be drawn into the pipette by capillary action.
2) Use 1 mL of a volatile solvent, such as ether, ethyl acetate, and pentane, to rinse this into a vial.
3) Dip a pipette's tip into this liquid.
4) Use 1 mL of the same solvent to rinse this through a pipe filter and into another vial.
5) It's time to inject your sample!

Solid sample preparation

1) Use 1 mL of one of the aforementioned volatile solvents to dissolve about 10 mg of the chemical.
2) Follow the preceding steps 3–5.

Operation of the GC

The GC can be run either with a temperature program (different temperatures) or at a single temperature (isothermal). The former is better when separating a small number of components, but for more intricate divisions, the latter is required to improve separation and quicken the process. Better separation of volatile chemicals and faster elution of less volatile molecules are possible with temperature programming. By reducing sample overloading, it also improves the separation process and overall efficiency of capillary columns.

It takes around 30 min to warm up the instrument. You should get your samples ready during this period. One person should use the GC while the other completes sample preparation as soon as the instrument is heated:

- Open the primary and shutdown valves on the gas tanks. Never change the regulator's pressure. It is recommended that He be at 52 psi, H_2 be at 20 psi, and air be at 30 psi.
- Uncover the column oven and note the column's parameters that you intend to use.
- Turn on the GC by flipping the switch on the rear right side.
- Adjust the oven temperature to 180 °C and all injectors and detectors to 250 °C.
- Place the flow meter's probe inside each column to determine the flow rate via that column.
- Note the rate of flow in each column. Moreover, gauge and note the split vent's flow rate. What counts is the percentage of your sample that really comes from the split vent flow divided by the column flow.
- Open the H_2 and air valves on the upper left side of each detector. After a short while, push the ignite buttons. Using a watch glass over the vent, you should be able to hear a tiny pop and verify that the flames are lit. On the cold glass, check for condensation. If absent, relight the lamp.
- Put the ATT at 2 and the range at 0. Verify that both detectors are turned on by checking the DET control.
- SIG 1 is linked to the data system. By pressing SIG 1, A or B, then ENTER, you can choose which column and detector to use.
- To initiate the ZERO operation, hit the ZERO button before each run and then press ENTER.
- Verify that PURGE B is turned on as well as PURGE A. The PURGE B valve should be set in the Timetable Events list to open at time 1.0 and close at time 0.0.
- This is required in order for samples to be injected correctly (using a syringe to inject the prepared samples might come after this).

Experiment 4: gas–liquid chromatography of pure liquids and mixtures

Objective
- To familiarize oneself with the principles of GLC.
- To acquire hands-on experience using a GLC and to discuss gas chromatograms.

Reagents and apparatus
GLC (preferably equipped with a FID and a digital integrator and column packed with SP containing 10% by weight of dinonyl phthalate), micro syringe, all GPR or pure liquid of ethyl acetate, octane, ethyl, *n*-propyl ketone, ketone and toluene, and mixture 1 (ethyl acetate: octane: ethyl) in unknown ratio.

Procedure
- Make mixture 2 (ethyl acetate: octane: ethyl), which has the constituents in equal weight proportions.
- Adjust the flow rate of the pure nitrogen carrier gas to 40–45 mL/min and the chromatograph oven to 75 °C.
- Once the oven temperature has reached a stable point, inject a 0.3 μL sample of combination 2 and determine whether each component's detector response is the same by looking at the peak areas.
- In the event that the detector response is different, replace the compound of an internal standard (*n*-propyl) with a weighted mixture of each of the individual components (ethyl acetate, octane, and ethyl).
- Inject a 0.1 μL sample of each mixture, measure the associated peak area, and then calculate the correction factors that will adjust the component peak areas (octane, ethyl, and ethyl) in relation to the internal standard (*n*-propyl).
- Make a weight-based mixture of the *n*-propyl and ethyl acetate.
- Inject a 0.3 μL sample of this mixture, assess the different peak regions, and calculate the percentage composition of mixture 1 after adjusting for variations in detector sensitivity.

Result
Record both the peak area and peak height of both samples and the internal standard (IS).

Calculation
Calculate the amounts of the samples

Experiment 5: elemental analysis using GC based on those adopted for the Carlo-Erba elemental analyzer

Objective
- To simultaneously ascertain the concentrations of carbon (C), hydrogen (H), and nitrogen (N) in samples that are organic and organometallic.

Apparatus and chemicals
Gas–liquid chromatography (preferably equipped with a thermal conductivity detector and a column of 2 m length), samples (about 1 mg), tin container, vertical quartz tube, standard compound (cyclohexanone-2,4-dinitrophenylhydrazone), Cr_2O_3, and Cu.

Procedure
- Weigh samples (typically 1 mg) in a tin container that is clean and dry. Samples are put into a vertical quartz tube that is kept at 1,030 °C in a steady stream of helium gas at predetermined intervals.
- Flash combustion happens when the samples are injected, momentarily enriching the helium (He) stream with pure oxygen.
- The resulting gas combination is passed over Cr_2O_3 to achieve quantitative combustion, and after that, it is passed over copper at 650 °C to eliminate surplus oxygen and convert nitrogen oxides to N_2.
- The gas combination is then finally run through a 2 m long, properly heated chromatographic column that is heated to about 100 °C.
- The N_2, CO_2, and H_2O components are separated and eluted to a thermal conductivity detector. An integrator, digital print out, and potentiometric recorder receive the detector signal in parallel.
- By burning standard compounds like cyclohexanone-2,4-dinitrophenylhydrazone, the device is calibrated.
- It is vital to discuss here the measurement of total organic carbon, which provides an estimate of the total amount of organic contaminants and is crucial for water analysis and quality monitoring. To extract carbon dioxide from any carbonate or hydrogencarbonate that may be present, the water is first acidified and then purged. Following this treatment, a precisely measured small volume of the water is injected into a gas stream. The gas stream next travels through a packed tube that is heated, oxidizing the organic material to carbon dioxide. The latter is measured using an FID in GC, or it can be converted to methane and measured using infrared absorption.

Determination of oxygen

To prevent oxidation, the sample is weighed into a silver container that has been cleaned with a solvent, dried at 400 °C, and stored in a closed container. It is dropped into a reactor that has been heated to 1,060 °C. A layer of carbon coated with nickel allows for the quantitative conversion of oxygen to carbon monoxide (see Note). The pyrolysis gases are then sent into a chromatographic column (1 m long) that is equipped with molecular sieves (5 cm) heated to 100 °C. A thermal conductivity detector is used to measure the CO after it has been separated from N_2, CH_4, and H_2.

Note: The addition of a chlorohydrocarbon vapor to the carrier gas is found to enhance the decomposition of the oxygen-containing compounds.

Determination of sulfur: The initial steps for the sample's flash combustion are basically the same as those outlined for determining C, H, and N. The combustion gases are then passed over tungsten (VI) oxide, WO_2, to achieve the quantitative conversion of sulfur to sulfur dioxide. Excess oxygen is then eliminated by passing the gases through a heated reduction tube that contains copper. The gas combination is then heated to 80 °C and sent through a Porapak chromatographic column, where SO is detected using a thermal conductivity detector and separated from other combustion gases.

1.5 High-performance liquid chromatography (HPLC)

HPLC is a type of liquid chromatography that is used to separate molecules that are dissolved in solution or MP, including polar and nonpolar compounds, PAH (inorganic ions), and nonvolatile materials. Analyses using a degasser and filtered sample solution can be performed at room temperature or lower. These days, compared to GC, HPLC is a larger and more important market. Both the fixed phase and the MP are separated according to the internal diameter of the column. The most widely used UV–vis detectors are wide-range, nondestructive, FG-sensitive, and three-dimensional.

It is distinguished by the high pressure used to force a MP solution through a SP column, enabling high-resolution separation of intricate combinations. As shown in the figure, HPLC instruments are made up of a detector, a separation column, a pump, an injector, and a MP reservoir. A sample combination is injected onto the column to separate the compounds. Two types of SPs are employed in HPLC: nonpolar C8, C18, and long-chain n-C18 hydrocarbon (octadecylsilyl group, or ODS) for the "reversed phase" and polar silica and alumina for the "normal phase." The choice of MP also depends on the phase's characteristics (for a normal phase, the most commonly used mobile.

Hexane, dichloromethane, isopropanol, and methanol make up the SP, and water, methanol, acetonitrile, and tetrahydrofuran (THF) make up the MP for the reverse phase. Due to variations in their partition behavior between the MP and the SP, the various components of the mixture pass through the column at different times. Degassing the MP is necessary to prevent air bubbles from forming and affecting the detector signal or peak.

The sample is transported into the column classified for isocratic elution, which uses a single solvent separation approach, and gradient elution, which uses two or more solvents that are changed throughout the separation process, using the solvent reservoir (MP). The chromatograph's brain, the HPLC pump, is utilized to provide the right pressure needed to force solvent into the sample. A pump that can move solvent at flows of up to 10 mL/min and pressures of up to 4,000 psi. The samples were injected using a syringe/injector (manual or autoinjector) or sample valve (fixed-volume loop of 1–200 μL).

Although heavy-walled glass tubing is infrequently found, smooth-bore stainless steel tubing is typically used to produce liquid-chromatographic columns. The latter can only be used at pressures less than roughly 600 psi. It is the central component of the chromatograph that offers the ability to divide a mixture into its constituent parts. The length of liquid chromatographic columns can vary from 10 to 30 cm, while their internal diameter typically ranges from 4 to 10 mm. The packing (alumina, Celite, and silica gel) has a particle size of 5 or 10 μm. Right now, the most widely used column is 25 cm long, 4.6 mm in internal diameter, and filled with 5 μm particles. These columns have between 40,000 and 60,000 plates/m (see figure 7).

There are three main types: analytical columns, which separate the sample components; guard columns, which are positioned before and after the injection port to extend the analytical column's useful life by eliminating contaminants and particulate matter from the solvents; and scavenger columns, which are positioned between the pump and injection valve and are used to saturate the aqueous MP with silica to lessen the impact of high pH (>8) buffers on the packing in the analytical column.

There are three main types: analytical columns, which separate the sample components; guard columns, which are positioned before and after the injection port to extend the analytical column's useful life by eliminating contaminants and particulate matter from the solvents; and scavenger columns, which are positioned between the pump and injection valve and are used to saturate the aqueous MP with silica to lessen the impact of high pH (>8) buffers on the packing in the analytical column.

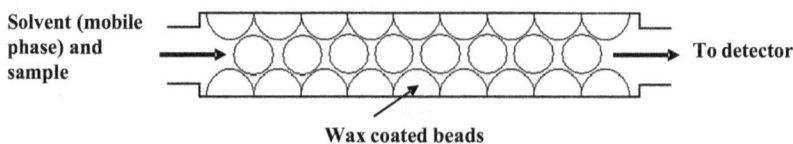

Wax coated beads

Similar to GC, HPLC offers a large range of detectors that can be either universal or specialized, destructive, or nondestructive, responsive to mass flow or concentration, and have much more difficult requirements when it comes to interacting with spectrometers in hyphenated procedures. Bulk property detectors are capable of responding to bulk properties of the MP, such as density, dielectric constant, or refractive index, while solute property detectors are capable of responding to properties of the solutes, such as diffusion current, UV absorbance, or fluorescence, that are not present in the MP.

Nuclear magnetic resonance, radiochemistry, conductometric, fluorescence, evaporative light scattering (ELSD), mass spectrometric (MSD), electrochemical (ECD), evaporative light scattering (LSD), photo diode array detectors (DAD), ultraviolet–visible (UV–Vis), refractive index (RI), fluorescence, and near-infrared (near-IR) are among the detectors used in HPLC. UV–Vis and fluorescence are the most prevalent types among those. The final piece of HPLC equipment is data processing software, which is connected to a HPLC system to receive data from the system and display it as a graph or chromatogram that shows quantitative data (area under curve) and qualitative data (retention time) (see figure 6).

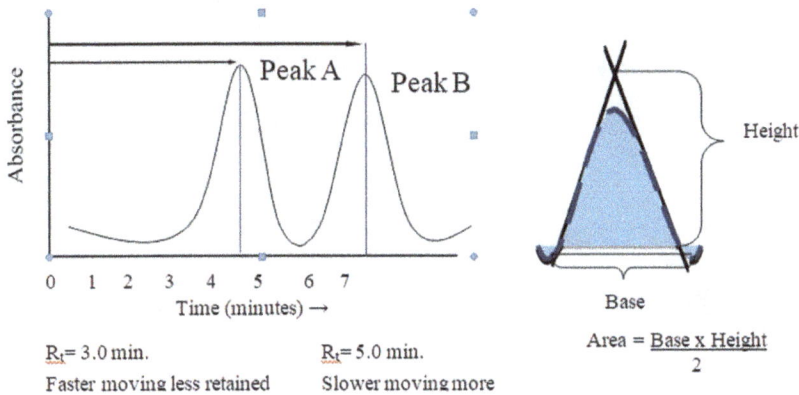

$R_t = 3.0$ min.
Faster moving less retained

$R_t = 5.0$ min.
Slower moving more

$Area = \dfrac{Base \times Height}{2}$

Figure 6: Chromatogram that describes qualitative data (retention time) and quantitative data (area under curve).

External calibration

$$\text{Calibration factor (CF)} = \frac{\text{Peak area (or height) of the compound in the standard}}{\text{Mass of the compound injected (in nanograms)}}$$

Internal calibration

$$\text{Relative retention time (RRT)} = \frac{\text{Retention time of the analyte}}{\text{Retention time of the internal standard}}$$

For each of the initial calibration standards, the RF values for each target compound relative to one of the ISs are as follows:

$$R_f = \frac{C_{is} \times A_s}{A_{is} \times C_s}$$

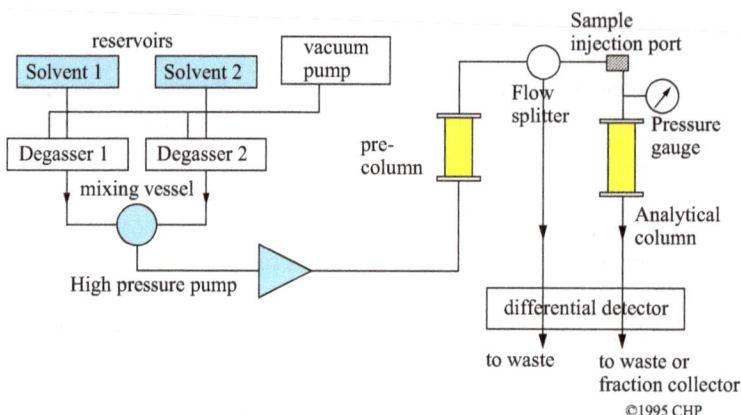

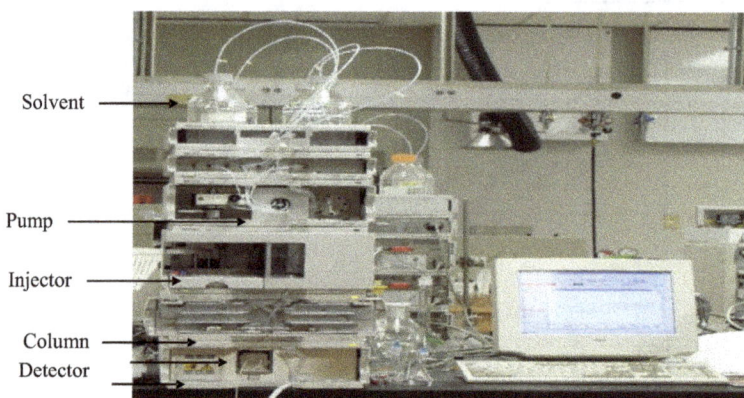

Figure 7: Instrumentation of high-performance liquid chromatography (HPLC) calibration of HPLC.

where A_s is the peak area (or height) of the analyte or surrogate, A_{is} the peak area (or height) of the IS, C_s the mass of the analyte or surrogate in the sample aliquot introduced into the instrument in ng, and C_{is} the mass of the IS in the sample aliquot introduced into the instrument in ng.

HPLC system affects
–   The SP, coating material, deactivation, and column material are the parameters of the column.
–   Temperature, flow, signal, sample sensitivity, and detector are the parameters of the instrument.
–   Concentration, matrix solvent, effect, and sample effect are the sample parameters. Efficiency, resolution, inertness, retention index, capacity factor, and column bleeds are used to evaluate those.

The factors that influence HPLC performance are
- the sample size;
- the polarity of the sample;
- solvent, and column;
- the pump pressure;
- the internal diameter of the column (the smaller the diameter;
- the higher the sensitivity);
- the temperature (the higher the temperature, the higher the separation).

Application
- The pharmaceutical sector regulates medication stability and determines drug quantity from pharmaceutical dosage forms.
- For example, Panadol tablet's paracetamol content and from biological fluids (e.g., blood glucose level).
- Examination of naturally occurring pollution (mercury and phenol in seawater).

1. The process of degassing in the mobility stages: The diagram illustrates a system intended to helium gas sparge four distinct reservoirs. It is sent to metal or fritted glass filters via a set divider switch, which drives out (sparges) additional dissolved pollutants by dispersing the air into an insignificant bubble fog. An alternative method for removing dissolved gases from a mixture would be to manually filter the mixture under vacuum before adding it to the reservoirs.

2. MP storage: There are two or more repositories for substances, one for every between two and four pristine liquids (water, acetonitrile, methanol, THF, etc.), possibly with a pH-controlling buffer. Take note of the inlet filter usage. As an alternative, the MP mixture might be manually prepared to the required composition and kept in a single reservoir. Isomatic HPLC elution refers to the process of operating with an arrangement that has just one persistent phase of mobility.

3. MP mixing: To create the appropriate mixture composition and combine liquids within the storage tanks, one can adjust the solvent distributing gate. For an isocratic operation, this mixture may have a fixed proportion, or the mixture proportions could be changed over time to alter the solvent strength of the MP. For a basic illustration, the parts of the MP of mobility might be set to change at time zero. At the end of the run, the ratio of 25% acetonitrile to 75% water was changed. Gradient elution is the phrase used to describe this change in the solvent strength or polarity of the MP during a run. The composition and elution strength of the MP change over time in a "gradient" manner. Thus, it is possible to separate and elute a mixture of components with widely varying polarity more quickly. In an RP division, less-polar analytes are going to dissolve slower in this gradient, but the process will be completed more quickly than in an isocratic one. Gradient elution, the HPLC counterpart of a thermostatic GC,

has all the same advantages. Should "slowly eluters" be required to be eliminated before the subsequent sample injection, it may be better to use back flush valves on both ends of the column to reverse the flow through the column and back flush these off the front of the column and out a drain port that's located between the column and the injector (not shown here) if these are very slow eluters that don't need to be measured. Re-establishing equilibrium between the SP and the original MP content is a laborious process after a gradient run. Isocratic elution is recommended unless the components to be eluted have a broad polarity range. It is easier to put into practice, and it will allow for lower analysis cycle times provided that the separated components' Rt values do not grow excessively long and be eliminated before the subsequent sample injection; that information is extremely significant.

4. HPLC pump: The reciprocating piston pump is a popular kind of HPLC pump (see figure 8). In the insertion cerebrovascular accident, the MP liquid is pulled from the solvent side, and during the stroke that produces exhaust, it is forced into the nozzle and to the top of the column. This is the source of the high pressure needed to enable HPLC's excellent performance. During the fill stroke, the pump's flow will decrease, and it will increase during the exhaust stroke. The downstream "pulse dampener" aims to simulate a steady, constant flow by mitigating flow surges. Repetitive pressure spikes would destabilize the SP particle bed. A change in flow would be reflected in a change in the response of mass-flow rate detectors. These are two negative effects of pulsed flow. By arranging two pistons in a single arrangement, improved engineering of pumps results in greater dependability of pressure as well as flow. In other words, one is in the fill stroke and the other is in the exhaust stroke.

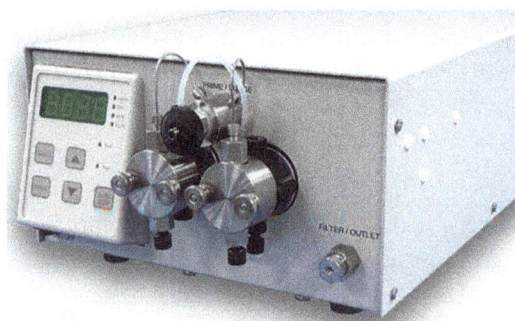

Figure 8: HPLC pump.

5. **Fill/drain valve**: There are many liquid transfer lines and components that the MP travels through a number of liquid transfer lines and other parts before reaching the column's head. All of this must be filled by the analyst with each fresh MP, and instead of using a complicated fluid stage containing salts in solution, buffering agents, and other ingredients, one should keep the filter column and arrangement in fresh-

water or a basic liquid after use. Additionally, it can be utilized to move the storage solvent or MP downstream in order to remove those.

6. Rotary sample **loop injector:** The most common design of HPLC injector is the loop injector. Loop injectors are the most widely used type of HPLC injector. It is not practicable to insert the sample into a liquid movement of the mobile component under extreme pressure using a syringe and a septum, as is done in GC. The button, which protrudes from the surface of the paper on an axis, is schematically shown in an end-on view at the top of the picture. Three outlet fittings and three inlet fittings are present. In the "load" position, the valve is visible. A syringe is used to inject a sample into a sample loop of tubing with a predetermined amount (such as 50 or 100 µL), with any extra sample going to waste. The valve is turned 60° clockwise to inject. The dotted lines in the valve schematic now correspond to the solid-line flow channels.

7. The column: It is easy to cut off and reattach a fused silica capillary GC column that has immobile pollutants accumulated at its front end. HPLC columns do not allow for this. Over time, even with the most meticulous filtration, residues of particulate matter may clog the small pores. The workaround involves attaching a very short, low-cost "guard column" in front of the "analytical column" that is frequently filled with "pellicular" particles – that is, particles that lack full internal porosity, which results in a significantly smaller bonded SP layer. When it becomes contaminated, it is taken out, disposed of, and changed with a brand-new guard column. If the stationary portion of the apparatus is capable of handling it, maintaining the column at greater temperatures may be advantageous.

Similar to GC, the column's diffusion and equilibration rates improve as temperature rises. In HPLC columns, the equilibrium of the entire column of temperature using a cooker requires longer than with capillary GC. Before the MP enters a thermostated HPLC column, it must be preheated to the required elution temperature in order to prevent zonal variations in temperature throughout the circumference of the column, which will impair clarity. Elution times can be more consistently achieved by thermostating an HPLC column, even in the absence of elevated temperature operation.

8. HPLC detectors: Similar to GC, HPLC offers a large range of detectors that can be either general or specialized, damaging or not, responsive to mass movement or focus, and have much more demanding specifications when it comes to interacting with spectrum analyzers within bracketed procedures. Connecting the waste product tip of the column to the detection sensor requires piping with an even lesser dimension than that specified in the subsection to avoid the consequences of extra-column band broadening. Stainless steel can be substituted with PEEK plastic, which has a higher density and strength and lower pressure or particular, damage or not, mass circulation or intensity receptive, with significantly more difficult specifications for punctuated methods' interface to spectroscopy instruments. To prevent extra-column band widening effects, tubing with an even narrower diameter than that mentioned previously in item 5 is required to

connect the effluent end of the column to the detector. It can be substituted with PEEK plastic, which has a higher density and strength and a reduced pressure.

9. A tiny capacity "movement cell," where the instrument for assessment is located and the column discharge passes through, is a feature shared by many HPLC detectors. Even detector response data files with complete spectrum information obtained from within a flow cell are occasionally not categorized as hyphenated HPLC procedures. Typically, HPLC consumption rates start at 1 mL/min. Approximately once per second or two, samples of the amount of analytes in the discharge may have been taken. Within this particular dimension, the wastewater volume has to pass via the circulation unit. Therefore, the assessment of circulation in a cell's quantities needs to be between 5 and 50 mL. Translucent panes at the entry and departure points are essential to flow cells housing HPLC detectors, as many of these devices depend on the release or uptake of radiation. The length of radiation routes is limited by the demand for tiny volumes. This is among the variables that determine how much optical spectroscopic signal may be obtained at a particular analyte concentration.

HPLC operational for experiments

Purge gas supply: Helium is used to purge the solvents for 15 min. Open the main valve and needle valve on the gas cylinder. Switch on the purge valve control. Adjust purge rates using knobs A, B, C, and D.

Solvent reservoirs: Bottle A contains water, Bottle B contains acetonitrile, Bottle C contains methanol, and Bottle D contains water buffered at pH 3.1.

Sample injector: The injector is a six-port rotary valve consisting of a 20 μL sample loop. Good sample injection procedures are:
1. Fill the 0.025 mL syringe (note that an HPLC syringe needle is blunted) with the sample and empty it into the waste about three times to clean the syringe.
2. Immerse the needle into the sample, and if necessary, pump the syringe by slowly aspirating and rapidly expelling the solvent to remove any air bubbles.
3. Fill the syringe to more than 20 μL.
4. Switch to the "Load" position. Insert the needle fully (you should feel some resistance) into the injection port and depress the plunger quickly.
5. Do not bend a needle or a plunger!
6. Make sure the "Not Ready" LED is off. Switch to the "Inject" position.

Column: Spherisorb ODS 2, 125 × 4 mm, 5 μm, 80 Å pore diameter, 0.5 mL/g pore volume, 220 m^2/g surface area, 12.0% w/w C loading, pH range 2–8, maximum pressure 6,000 psi.

Detector: An UV-absorbance detector is used. Set the wavelength to 254 nm.

HPLC startup procedure
–   Turn on the switches on the detector and the pump.
–   Slowly open the main cylinder valves for He for purging.
–   The HPLC method for this experiment is stored in the directory on the hard drive of the control computer.
–   Create your own subdirectory in the directory for the storage of your data, giving each data file a name unique to you and your partner (chem316labmanual, 2013).

Software control
–   In Run Control Setup, under Sample Info, enter the filename and comments.
–   In Instrument Setup, check the pump flow, input flow rate, and solvent composition. For isocratic separation, make sure the bottom gradient elution table is empty.
–   In Instrument|Setup VWD Signal, set the wavelength to 254 nm.
–   When you are ready to load the sample, run the Control Run method and watch for the "waiting for injection" message.
–   Then load your sample and inject it (chem316labmanual, 2013).

Experiment 6: quantification of caffeine in various drinks by HPLC

Objective
To measure the amount of caffeine in cola soft drinks, an IS is used, and sample clean-up methods employing solid phase extraction.

Theory
HPLC can be used in an isocratic setting to measure the caffeine content in soft drinks like Coke and Pepsi as well as liquids like coffee and tea. An IS is used in this process. When there has been sample preparation, chromatographic conditions are unstable, or the injection technique may not be exact, the IS method is employed. Since theophylline's structure and molecular characteristics are similar to those of caffeine, it is utilized as an IS in this context. Since 8-chlorotheophylline is somewhat more water-soluble than theophylline, it is another often-used IS used for this analysis.

Apparatus and chemicals
UV detectors (275 nm), C18 column, methanol/water (35:65) at a flow rate of 1 mL/min, 250 µL injection loop (actually, a 10 µL loop is used and only 10 µL goes onto the column), extracted and filtered theophylline IS (10 ppm), standard filtered caffeine solution (10 ppm), 0.45 µm filter, HPLC grade methanol, and deionized water.

Procedure
1. Filtered coffee (10 ppm). This solution is used to calculate the caffeine retention time, which allows it to be distinguished from theophylline. Before injection, it needs to be filtered via a 0.45 µmfilter.
2. Filtered 10 ppm of theophylline and 10 ppm of caffeine. Using this solution, the proper chromatographic conditions for analysis are created, and the analysis is calibrated by determining the relative response factors (RRFs) for the IS and the analyte. Before injection, it needs to be filtered via a 0.45 µm filter. Determine caffeine's relative response to theophylline based on the peak areas.

Calculate the RRF for caffeine.

$$RRF = \frac{\text{Concentration of analyte} \times \text{peak area of internal standard}}{\text{Peak area of analyte} \times \text{concentration of internal standard}}$$

Procedure 3 (solution 3): 10 ppm caffeine and theophylline extracted by solid phase extraction (SPE)

Step 1: Condition the solid phase extraction column (C18) with 5 mL of methanol followed by 5 mL of deionized water. (Conditioning)

Step 2: Empty the eluent from the column and inject 3 mL of the calibration solution (solution 2). Theophylline and caffeine will stay on the column. (Filling up)

Step 3: Use 5 mL of deionized water to flush the solid phase column. (Cleaning or washing)

Step 4: After attaching a 0.45 µm filter to the bottom of the C18 extraction cartridge, 3 mL of HPLC-grade methanol are injected into the cartridge to extract the caffeine and theophylline from the column into a sample tube. After that, this solution can be dried and reconstituted, but not before being thoroughly examined. (Removing)

Step 5: A second use for the C18 extraction cartridge will occur. To clean the cartridge, flush it with 3×5 mL of methanol and then 2×5 mL of water.

Step 6: After extraction and filtration (solution 3), inject 250 µL of the caffeine and theophylline solution and record the regions. The regions prior to extraction were identified beforehand. Using the following formula, find the percentage recovery for both theophylline and caffeine in the clean-up step:

$$\% \text{ Recovery } = \frac{\text{Peak area after extraction}}{\text{Peak area before extraction}} \times 100$$

Do the percentage recoveries vary? If so, describe how the IS technique is used.

Method 4 (solutions 4 and 5): 10 ppm theophylline was introduced to Coke and Pepsi, which was then extracted and filtered using SPE cleaning techniques.

The technical techniques remain the same as before, with the exception that the starting solutions (10 mL with 10 ppm added theophylline) for these extractions should be prepared using degassed soft drinks diluted 1 in 10. (NB: After sampling each soft drink, make sure to clean the cartridge as previously mentioned.) One by one, inject 250 µL of each soft drink's solution. Determine the caffeine content of the samples by utilizing the previously determined:

$$\text{RRF [caffeine]} = \frac{\text{Peak area caffeine RRF}}{\text{Peak area internal standard}}$$

Use the following format for recording your results and calculations in the practical record book.

Caffeine in coke
Conc. internal standard (theophylline) = 10 ppm
Peak area internal standard =
Peak area caffeine = RRF (caffeine) =
[Caffeine] in solution = show calculation
[Caffeine] in sample = (µg/mL) (NB: dilution factor!)

Discussion

1. Examine the variations in caffeine content between the soft drinks and the amounts documented in the literature.
2. Talk about using internal standards.
3. Talk about using SPE.
4. Go over the sample solutions' chromatograms.
5. What are the molecular weight and structure of caffeine?

Experiment 7: analysis of four types of acids in tablets by HPLC

Objective
- To investigate how the composition of the MP affects isocratic separation and to devise an optimal gradient elution scheme.
- To use the standard addition procedure and IS to perform a quantitative analysis of an unknown.

Apparatus and chemicals
HPLC-grade acetonitrile and water, acidic aqueous buffer (containing K_2HPO_4 adjusted by H_3PO_4 to pH 3.10), 100 ppm of phthalic acid (50 mL × 10 = 500 mL), 100 ppm of salicylic acid (50 mL × 10 = 500 mL), 100 ppm of benzoic acid (50 mL × 10 = 500 mL), 100 ppm of *p*-nitrophenyl acetic acid (100 mL × 10 = 1,000 mL), seven 50 mL volumetric flasks, and four 10 mL graduated pipettes.

Procedure
1. Impact of mobile phase composition on the separation in a 100 mL flask using a water/acetonitrile mobile mixture. Make a 50 mL test mixture with 25 ppm of each ingredient.

(A) Isocratic distancing: Utilizing a MP of water/acetonitrile in the ratios of 60:40, 70:30, 80:20, and 90:10, run the test mixture at 1.5 mL/min. Determine the sequence of elution by using distinct standard solutions.

Q1. Determine each compound's retention factor, k.

Q2: Determine the resolution, or R_s, of each chromatographic run's separation of two closely eluted molecules.

Q3: Determine the water/acetonitrile blend that yields the best compromise between retention time and resolution.

Q4: Has the composition of the MP affected the solutes' elution order? Describe in terms of the shift in the polarity of the MP.

Q5: Determine the optimal solvent blend's polarity index.

Q6: Determine the solvent composition of water/methanol needed to provide the same polarity index as previously obtained if methanol were to be employed in place of acetonitrile.

Gradient elution
Utilizing your understanding of isocratic separations with different water/acetonitrile compositions, run a gradient elution program to achieve the best solute separation resolution in the shortest amount of time.

Q1. Compare and contrast the results of gradient and isocratic elution methods for the solute separation in the test mixture, providing numerical examples.

Isocratic separations using various water/acetonitrile compositions perform a gradient elution program to achieve the best separation resolution of the solutes in the shortest time. Compare the resolution of isocratic versus gradient elution for the solute separation in the test mixture, providing calculations. Perform isocratic separations using different water/acetonitrile compositions. Run a gradient elution program to achieve the best separation resolution of the solutes in the shortest amount of time.

1. Quantitative analysis of an "unknown" sample by the standard addition method (six 50 mL flasks).

In any chemical analysis, there are possibly some changes in the analytical signal due to the presence of interferences or the matrix. This is called interference or, in general, the matrix effect. The standard addition method is well suited to minimize any errors caused by the matrix effect (as you have learned in atomic spectroscopy). In the case of two unresolved or partially resolved peaks in chromatography, one analyte can be considered an interferent to the other if we take the combined peak area as the analytical signal. Therefore, from your previous experiments, a 60:40 composition does not result in two peaks completely resolved from each other. Imagine that you could not improve the resolution, then use these separation conditions to illustrate the concept of the matrix effect and the use of standard addition to minimize errors. Prepare an "unknown" sample (50 mL) that contains the analyte of your choice (say, *p*-nitrophenylacetic acid at 25 ppm) and the partially resolved substance (at 25 ppm). Any one of the other two substances (at 25 ppm) can be used as the IS. Use 25 mL of your test mixture as the unknown sample; perform the standard addition method to establish the concentrations of the added analyte to be 0, 10, 20, 30, and 40 ppm, using the 100-ppm stock solution of the analyte. Remember to make up the final volume to 50 mL using distilled water.

Q1: To produce 0, 10, 20, 30, and 40 ppm of additional analyte in each example, calculate the volumes of 100 ppm analyte to be added to each of the five samples. Next, execute an HPLC analysis on each of the five samples.

Q2: Using the combined peak areas of the two overlapping peaks in each case, determine the peak area ratios for the five samples by using an internal reference. (Remember that you'll utilize the IS to reduce sample loss errors and standard addition process to reduce matrix effect mistakes.)

Q3: Calculate the analyte concentration in the "unknown" sample using linear regression analysis. Additionally, estimate the related uncertainty. As you wait for the results, set up the spreadsheet.

Q4: Discuss the benefits and drawbacks of the conventional addition technique (chem316labmanual, 2013).

2 Electrophoresis

General theory

Based on the differential migration of charged substances in a semiconductive media or solutions under the influence of an electric field, electrophoresis is a separation technique. The first two to study it were O. Lodge (1886) and F. Kohlrausch (1897), who reported on the movement of hydronium ions (H^+) in phenolphthalein gel and saline solutions, respectively. In its traditional form, electrophoresis uses an applied electric field that produces a potential gradient to differentially migrate mixtures of charged solute species or particles through a buffered electrolyte solution supported by a thin slab or short column of a polymeric gel, such as agarose or polyacrylamide.

The electrolyte, which is held in reservoirs at opposite ends of the supporting medium, is in contact with two platinum electrodes (the cathode and anode), which are connected to an external DC (direct current) power source.

When separations occur at higher operating voltages, significant heat may be produced (Joule heating), and many systems use water cooling to keep the operating temperature stable. The following outcomes are possible when the electrophoretic medium is heated:

(1) A faster rate of diffusion between the sample and the buffer ions, which causes the separated samples to enlarge.
(2) Samples that are highly susceptible to heat experience thermal instability. The two examples are enzyme activity loss and denaturation of proteins.
(3) A drop in buffer viscosity, which results in a drop in the medium's resistance.

Macromolecules in electrophoresis are defined by how quickly they move in an electric field, and the separation of big molecules (macro) is dependent on two forces: mass and charge.

These two forces come into play when biological material, like proteins or DNA, is combined with a buffer solution and put to a gel. By acting as a "molecular sieve," the gel substance divides the molecules into different sizes. When the electrical current is delivered, molecules are compelled to pass through the pores. The molecules are simultaneously attracted to the other electrode and repelled by the electrical current flowing through one electrode.

The following factors affect the rate of migration through the electric field:

(1) strength of the field,
(2) size and shape of the molecules,
(3) ionic strength and temperature of the buffer in which the molecules are moving.
   Buffer solutions undergo electrolysis, producing hydrogen (at the cathode) and oxygen (at the anode) and the reservoirs have to be replenished or the buffer renewed to maintain pH stability.

https://doi.org/10.1515/9783111575636-003

Additionally, a frictional barrier hinders this charged molecule's motion. This force of friction is a measurement of the:
(1) molecule's hydrodynamic size,
(2) molecule's form,
(3) medium's pore size, and
(4) buffer's viscosity.

Electrophoresis

The classification of electrophoresis is frequently based on the existence or nonexistence of a solid matrix or supporting medium that allows charged molecules to flow through the electrophoretic system. When a solid support medium is not available, aqueous buffers are used in solution electrophoresis systems. Due to the charged molecules' tendency to diffuse, these systems may experience sample mixing, which can lead to a loss of resolution during the sample application, separation, and removal processes. As a result, some method of stabilizing the aqueous solutions in the electrophoresis cell must be used by solution electrophoresis systems. Soluble-gradient electrophoresis systems, for instance, reduce diffusional mixing of the materials being separated during electrophoresis by employing different densities of a nonionic solute such as sucrose or glycerol (Figure 9).

The majority of practical uses for electrophoresis in biochemistry involve zonal electrophoresis, where samples are applied as spots or bands of material, and the aqueous ionic solution is conveyed in a solid substrate. Systems for zonal electrophoresis include paper electrophoresis, gel electrophoresis, and cellulose acetate and cellulose nitrate strips. The single set of general principles that govern all forms of electrophoresis is represented by the following equation:

$$\text{Mobility of molecule} = \frac{(\text{Applied voltage})(\text{net chrge on the molecule})}{\text{Fraction of the molecule}}$$

The mobility or rate of migration of a molecule increases with increased applied voltage and increased net charge on the molecule. In contrast, when molecular size and shape grow, molecular friction – or the resistance to flow in the viscous medium – increases and a molecule's mobility decreases. Given that mobility is the rate of migration, the total actual movement of the molecules increases with increasing time. Charge and frictional coefficient, which are based on the physical characteristics of molecules, are factors that affect mobility.

Additionally, temperature plays a role; the heating of the separation matrix needs to be managed. In order to maintain a constant temperature during electrophoresis, it will be necessary to offer a means of cooling the apparatus. It also relies on the particle's size, shape, and applied voltage. Consequently, "mobility" provides details regarding the charge, molecular size, and molecular structure.

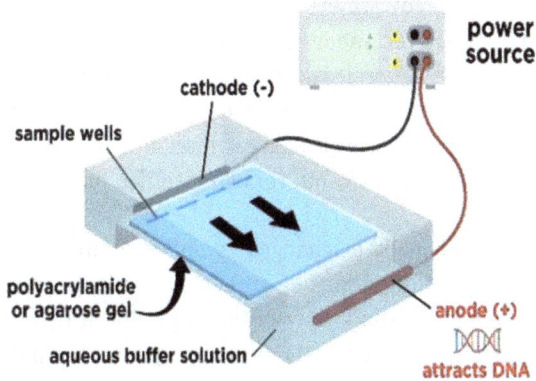

Figure 9: A solution electrophoresis system.

All of the cross-sectional regions of the paper strips, gels, or liquids used in the electrophoretic separation are typically subjected to an equal and constant voltage in electrophoretic systems. The best way to describe these electric fields is in terms of volts per linear centimeter. Nonetheless, voltage (V) is a function of current (I) and resistance (R), according to Ohm's law ($V = IR$). The resistance in the system is determined by the buffer's composition and the type of electrophoresis device. The system's resistance is crucial since it will control how much heat is produced throughout the electrophoresis process.

The samples are added to a buffer solution, which acts as a pH stabilizer and an electrically conducting medium for the duration of the separation. A layer of finely powdered materials (such as starch grains or diatomaceous earth), a sheet of paper, a gel (such as agar, silica, starch, or synthetic polymers), or a density gradient of non-ionized solutes are common ways to stabilize the solutions used in investigations.

The separation is determined solely by electrical mobility in cases when the stabilizing medium is only mildly sorptive. Electrical mobility, concentration, and sorbability of the migrating species dictate the separation in the event that the stabilizing medium is selectively sorptive. Solutes migrate through a solid medium made of a polymeric gel that holds the running buffer in place during certain types of electrophoresis. The characteristics of the particle itself, such as its charge, hydration, and dissociation tendency, as well as the qualities of the solution, such as pH, ionic strength, viscosity, and temperature, as well as the applied electric field strength and time, all affect the pace of migration.

For example, the ionized components of a mixture separate and migrate at various rates under a DC electric field. Since many solutes are weakly acidic, basic, or amphoteric, even little pH changes can have an impact on their mobility; hence, buffers play an essential role in electrophoresis. Amphoteric substances can have a positive, negative, or zero charge depending on the circumstances, but neutral species do not migrate and stay at or near the point at which the sample is applied. Cationic solute species, which are positively charged, migrate toward the cathode, and anionic species, which are negatively charged, migrate toward the anode. Each solute's electrophoretic mobility, m, which depends on its net charge, general size and shape, and the viscosity of the electrolyte, determines the rate of migration. The size and charge of each individual solute determines its mobility: cationic species are larger than anionic species in an ascending order of size, neutral species are next but remain together, and so on (see figure 10).

The selectivity of a separation can occasionally be changed by buffer additives: inorganic salts that are used to change the conformations of proteins; organic solvents that modify the electroosmotic flow, which is a bulk flow of liquid produced by hydrated buffer cations that tend to be drawn toward the cathode, when a potential gradient is applied across a running buffer; urea that denatures oligonucleotides and solubilizes proteins; surfactants that form micelles and cationic ones reverse charge on capillary wall; and cyclodextrins that provide chiral selectivity. For pH ranges in electrophoresis, running buffers such as MES (2-4-morpholinoethanesulfonic acid, 5.2–7.2) and TRIS (2,3-dibromopropylphosphate, 7.3–9.3) are employed. Phosphate (1.1–3.1), ethanoate (3.8–5.8), phosphate (6.2–8.2), and borate (8.1–10.1) are also utilized.

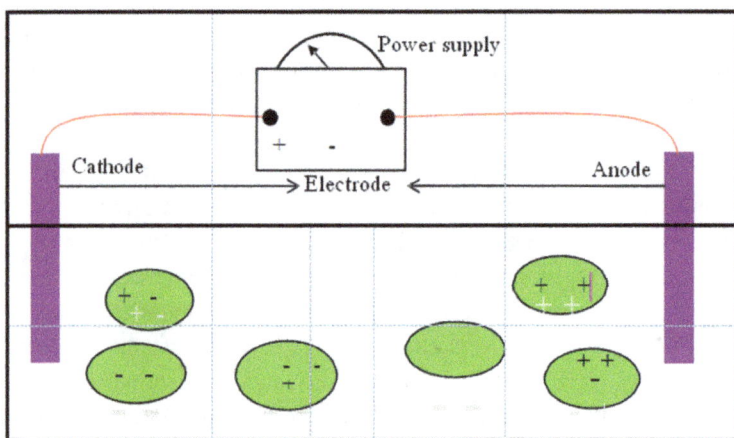

Figure 10: Instrumentation (setup) of electrophoresis. Marek Minárik, Analytical Science (Third Edition), 2019.

The application of a significant potential gradient over an extended period of time maximizes $\Delta d = (\mu_1 - \mu_2) \times t \times (E/S)$. However, lengthy separation durations should be avoided since, similar to chromatographic separations, band broadening caused by the solute species' diffusion in the buffer solution negatively reduces resolution.

Velocity of the molecule V is the ratio of the product of electric field strength E, and net charge q, to fractional coefficient, f:

$$V = \frac{E \times q}{f}$$

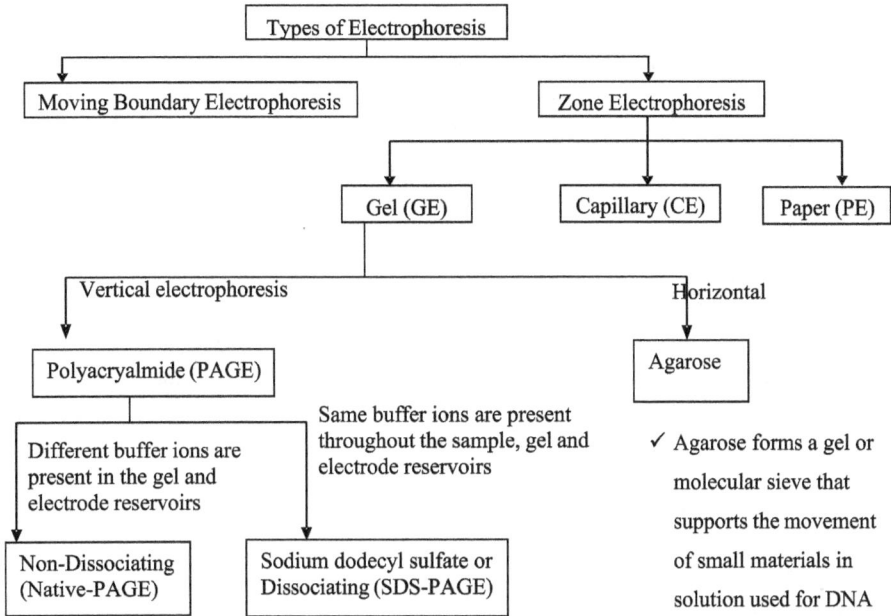

```
                    ┌─────────────────────────┐
                    │  Types of Electrophoresis │
                    └─────────────────────────┘
         ┌────────────────────────┴───────────────────────┐
┌──────────────────────────────┐              ┌──────────────────────┐
│ Moving Boundary Electrophoresis │           │  Zone Electrophoresis  │
└──────────────────────────────┘              └──────────────────────┘
                        ┌──────────────┬──────────────┬──────────────┐
                  ┌──────────┐   ┌──────────────┐  ┌──────────┐
                  │ Gel (GE) │   │ Capillary (CE) │ │ Paper (PE) │
                  └──────────┘   └──────────────┘  └──────────┘
```

Vertical electrophoresis   Horizontal

┌─────────────────────────────┐   ┌──────────┐
│ Polyacryalmide (PAGE)   │   │ Agarose │
└─────────────────────────────┘   └──────────┘

Different buffer ions are | Same buffer ions are present throughout the sample, gel and electrode reservoirs | ✓ Agarose forms a gel or molecular sieve that supports the movement of small materials in solution used for DNA
present in the gel and
electrode reservoirs

┌──────────────────┐   ┌──────────────────────────┐
│ Non-Dissociating │   │ Sodium dodecyl sulfate or │
│ (Native-PAGE)   │   │ Dissociating (SDS-PAGE)   │
└──────────────────┘   └──────────────────────────┘

✓ Made of Polyacrylamide

✓ Used for Protein molecular size, shape, charge

✓ IEF electrophoresis

✓ Western Blot technique

❖ Gel ingredients includes: Acrylamide, bis acrylamide, tris-HCl, N,N,N",N"-tetramethylethelenediamine (TEMED) and ammonium per sulfate (APS)

Paper electrophoresis

For the study and resolution of tiny molecules, paper electrophoresis is a widely used electrophoretic technique. As the adsorption and surface tension involved with paper electrophoresis typically modify or denature macromolecules, leading to poor resolution, this approach is not utilized to resolve macromolecules, such as proteins. Samples are usually applied to the electrophoretic paper using one of two approaches. A sample of solutes dissolved in distilled water or a volatile buffer is applied as a tiny spot or thin stripe on a penciled "origin line" on the paper in the dry application technique. At other points along the origin line, recognized compounds with the appropriate standards are used. The origin line should be in the middle of the paper if electrophoretic migration toward both system poles is anticipated. The origin line should be close to one end of the paper if you only expect one direction of migration. After the solvent containing the samples has evaporated, the paper is dampened with the electrophoresis buffer, either by uniform spraying or by dipping and blotting the ends of the paper so that wetting of the paper from both ends meets at the origin line simultaneously.

In the **wet** application procedure, samples dissolved as concentrated solutions in distilled water are applied to paper predampened with electrophoresis buffer. The dry application procedure has the advantage of allowing small initial sample spots and better resolution of similarly mobile compounds. However, this method is awkward because the dipping or spraying requires considerable skill to avoid spreading the applied samples. The wet application procedure is simpler to perform, but usually yields larger spots and poorer resolution because of sample diffusion.

The paper is placed in the electrophoresis chamber with both ends in contact with the reservoirs of the electrophoresis buffer at the electrodes after the sample has been applied to the dampened piece of paper. The paper must be positioned to allow for maximal migration toward the appropriate electrode if the origin line is not in the middle of the sheet. An electric field is introduced into the system after the chamber has been sealed or covered to prevent electric shock. Heat is produced by the application of the electric field and the ensuing resistance to the current flow in the buffered paper. This is the biggest challenge when using paper electrophoresis.

Heat dries the paper, which in turn leads to more resistance to current flow, which causes greater resistance, and so forth. Most low-voltage systems are portable and can be operated in cold rooms or refrigerated chambers. In contrast, most high-voltage systems either employ a cooled flat-bed system to dissipate the heat or operate in a cooled bath of inert and nonpolar solvent (e.g., Varsol, a petroleum distillate). This solvent absorbs the heat generated by the system without mixing with the water, buffers, or samples on the paper.

The presence and position of the molecules of interest are identified after the samples are electrophoretically resolved on the paper for the length of time and voltage needed for the best separation. The current is then cut off, the paper is taken out, and dried.

Capillary electrophoresis

Capillary electrophoresis (CE) is a type of gel electrophoresis in which the solutes to be analyzed are placed in a long, fine-bore capillary tube, usually 50–100 cm long and 25–100 m inside diameter. The electrophoresis medium, which is normally aqueous, is used in this process. At one end of the capillary, a very little sample – nanoliters or less – is deposited and electrophoresed at voltages of up to 20–30 kV. Using any of the techniques frequently employed in high-performance liquid chromatography, the analytes are monitored when they come from the opposite end of the capillary, CE offers the advantages of extremely high resolution, high speed, and high sensitivity for the analysis of extremely small samples, but is obviously not useful as a preparative method. It has proven especially useful in the separation of DNA molecules that differ in size by as little as only a single nucleotide.

In some of the most recent DNA sequence designs, polynucleotide separation is used on CE due to its high resolution. The separation of uncharged molecules can also be accomplished with CE by incorporating charged detergent micelles, such as sodium dodecyl sulfate (SDS), into the aqueous electrophoresis medium. Electrophoresis can be used to separate a mixture of solute molecules that partition between the hydrophobic interior of the micelles and the aqueous medium when added to such a system. CE is a very flexible technique, and research is constantly being done to determine the best approach and range of applications.

Gel electrophoresis

In gel electrophoresis, molecules are separated in aqueous buffers supported within a polymeric gel matrix. Molecules are separated by size and charge using gel electrophoresis. Because they are in an electrical field, the charged macromolecules move over a gel. In accordance with the size and form of the molecules, the gel functions as a sieve to delay their passage. It is employed in the nucleic acid and protein separation process. A variety of gels, including agarose, polyacrylamide, and starch, are utilized as supporting media. Starch was used in the earliest gel electrophoresis experiments. It offered the initial proof of isozyme existence. For proteins, SDS–polyacrylamide gels (often SDS–PAGE) are utilized, and for nucleic acids, agarose. Systems for gel electrophoresis offer a number of noteworthy benefits. They can be utilized for preparative-scale electrophoresis of macromolecules because, firstly, they can handle larger samples than the majority of paper electrophoresis systems. Secondly, the gel matrix's characteristics can be changed to suit certain applications. The gel increases the friction that controls the electrophoretic mobility, making this possible. Polymerized gel systems with minimal frictional resistance to macromolecule migration can be primarily used as stabilizing or anticonvection devices due to low concentrations of matrix material or low levels of monomer cross-linking. Alternatively, more friction is produced, leading to molecular sieving, by using higher quantities of matrix material or a higher degree of monomer cross-linking.

Molecular sieving is a situation in which viscosity and pore size largely define electrophoretic mobility and migration of solutes. As a result, the migration of macromolecules such as nucleic acids and lipoproteins in the system will be substantially determined by molecular weight. Agarose (a polygalactose polymer) gels have proven quite successful, particularly when applied to very large macromolecules. PAGE are formed as a result of polymerization of acrylamide (monomer) and N,N-methylene-bisacrylamide (cross-linker), which is one of the most useful and most versatile in gel electrophoretic separations because they readily resolve a wide array of proteins and nucleic acids. The acrylamide monomer and cross-linker are stable by themselves or mixed in solution, but polymerize readily in the presence of a free-radical-generating system. The free radical initiator APS and an N,N,N',N'-tetramethylethylenediamine (TEMED) catalyst are added in the chemical approach, which is the most widely employed (table 3). These two elements produce the free radicals required to initiate polymerization when combined with the monomer, cross-linker, and suitable buffer. APS is substituted in the less common photochemical approach with a photosensitive substance, such as riboflavin, which releases free radicals when exposed to UV light. You could reduce the amount of cross-linker and/or monomer in the polymerization solution if bigger pores are needed. One can increase the concentration of cross-linker and/or monomer if smaller holes are needed.

Table 3: Effective separation range and recipe for polyacrylamide gels with various percent acrylamide monomers for use with sodium dodecyl sulfate–polyacrylamide gel electrophoresis (SDS-PAGE). Sambrook, J., Fritsch, E.F., and Maniatis, T. (1989)

Component or recipe (mL)	% Acrylamide in resolving gel					
	4	7.5	10	12	15	20
Distilled water	2.7	9.6	7.9	6.6	4.6	2.7
30% Acrylamide solution	0.67	5.0	6.7	8.0	10.0	11.9
1.5 M Tris chloride (pH 8.8) and (pH 6.8)	0.5	5.0	5.0	5.0	5.0	5.0
10% (w/v) SDS	0.04	0.2	0.2	0.2	0.2	0.2
TEMED	0.004	0.008	0.008	0.008	0.008	0.008
10% (w/v) Ammonium persulfate	0.04	0.2	0.2	0.2	0.2	0.2
Effective separation range (Da)		45,000–200,000	20,000–200,000	14,000–70,000	5,000–70,000	5,000–45,000

Another technique for separation that combines elements of HPLC and electrophoresis is electrochromatography. The procedure for figuring out the molecular weight of an unidentified nucleic acid sample is precisely the same as the one explained for figuring out the molecular weight of proteins in SDS-PAGE.

Electrophoresis of samples:
– Samples: boiled 3 min with loading dye (2× Laemmli buffer + running dye)
– Mini-PROTEAN tetra cell: set up according to the SOP given in workbook
– Power settings: 75 V for 45–60 min
– Running dye should not run off the bottom of gel

As a result, the use of electrophoresis encompasses the following tasks: partial substance descriptions, separation of mixes, isolation of mixture components, detection and identification of specific compounds, and quantitative measurements. Further improvements to the method's efficiency led to the development of well-known procedures like gel and paper electrophoresis. Paper electrophoresis gained significant success in the 1950s, but it is now seen as outdated. In contrast, gel electrophoresis is still frequently used, particularly in biochemistry, to determine the presence of proteins and nucleic acids.

Process of electrophoresis

–   Sample application
–   Adjustment of voltage or current DC! (gel electrophoresis about 70–100 V)
–   Separation time: minutes (e.g., gel electrophoresis of serum proteins, 30 min)
–   Electrophoresis in supporting medium: fixation, staining, and destaining
–   Evaluation: qualitative (standards) and quantitative (densitometry)

For example, serum protein electrophoresis separation procedure:
    Sample preparation → Gel loading → Gel running → Protein staining → Quantifications

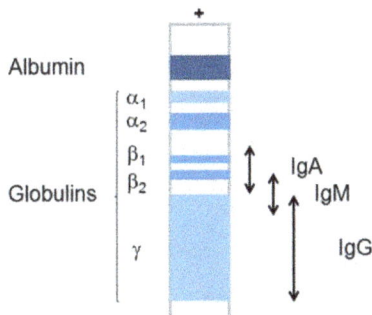

Types of protein electrophoresis	Standard or reference ranges
Total protein	6.0–8.0 g/dL
Albumin	3.5–5.0 g/dL
α1-Globulins	0.1–0.4 g/dL
α2-Globulins	0.4–1.3 g/dL
β-Globulins	0.6–1.3 g/dL
γ-Globulins	0.6–1.5 g/dL

Generally, this method is used to (1) distinguish molecules on the basis of charge and shape, (2) to determine the molecular weight of proteins, (3) to identify the transitions between charged and uncharged residues in amino acids, and (4) to quantitatively differentiate various molecular types. Over the years, electromigration methods (origins, concepts, and applications) have made a substantial contribution to analytical chemistry and biopharmaceutical science. Electrophoresis remains the method of choice for the study of proteins, amino acids, and DNA fragments, particularly in the field of biochemistry.

CE has allowed the technique to develop into a high-performance instrumental approach. Applications range widely, including medicines and tiny organic and inorganic charged or neutral substances. At the moment, CE is regarded as a well-respected instrument available to analytical chemists for the resolution of numerous analytical issues. The fields of protein analysis and DNA sequencing continue to be the main application areas, along with low-molecular-weight molecules (pharmaceuticals). In fact, the completion of the human genome project several years ahead of schedule was made possible by CE.

3 Electroanalytical methods

Electroanalytical methods are methods based on the investigation of electrochemical phenomena.

Electroanalytical methods are widely used in chemical laboratories and industrial process controls. They offer qualitative investigation for things like an organic solvent's clarity and relative quantitative analysis for substance control, including comparing the total ion content of water that is consumed. The variables measured are electrochemical quantities of the system investigated, such as current, potential, resistance, and charges. The methods are applied for both quantitative and qualitative investigations. The concentration of a substance can be determined on the basis of measuring the electrode potential of a suitable indicator electrode (direct potentiometry), the charge flowing through a cell (coulometry), or the diffusion-controlled current (polarography). In this manual, only the above three methods are described, which the students will deal with during the laboratory course.

3.1 Conductometry

Numerous solutions can be subjected to conductivity testing; sample preparation is not necessary. The technique is used to track the ionic concentration of solutions in drinkable beverages, natural water, deionized high-purity water, drinking water, and high-purity solvents. In ion chromatography, high-performance liquid chromatography, and other chromatographic methods that generate charged species, conductivity is frequently used as a detector. In both aquatic and nonaquatic solvents, the technique of titration using conductivity is an effective technique for endpoint detection.

An alternating current (AC) voltage is supplied between two electrodes submerged in the same solution during conductometry. There is a current flow because of the applied voltage. The solution's electrolytic conductivity determines the current's magnitude. While the measurement alone cannot identify the species carrying the current, this approach can detect changes in contents in a sample during chemical reactions (e.g., during a titration). Conductivity and conductance assessments are widely used to monitor the purity of streams used for processing and water. Commercial ion chromatography equipment uses conductivity detectors to measure ion concentrations. Platinum-plated electrodes are used to lessen the polarizing effect of current passage between the electrodes for accurate conductance measurements. The Pt conductors were prepared by electropositive a thin layer of Pt black from a 2% platinic chloride solution in 2 N HCl, and washing the Pt in warm, concentrated HNO_3 can be rendered platinized. For the intent of conducting conductometric titrations, the end point needs to be accurately found; however, the absolute conductance is not as crucial. Bright platinum, tungsten, and other electrodes can be employed instead.

https://doi.org/10.1515/9783111575636-004

According to Ohm's law, the equation can be used to determine the resistance of metal wire:

$$R = \frac{E}{I}$$

where R is the wire's resistance (V), I is the current of electrons passing through the wire (A), and E is the voltage applied to the wire (V).

The conductor's size determines the resistance:

$$R = \frac{pL}{A}$$

where p is the resistivity, L is the length, and A is the cross-sectional area. Another valuable parameter, especially when we consider the mechanisms of current flow in solutions, is electrolytic conductivity or specific conductance (k).

The units of electrolytic conductivity are $\Omega^{-1}\,m^{-1}$, also called $mhom^{-1}$, or S/m.

Actual units of measurement are typically expressed in mS/cm, but the SI equivalent is S/m. An intrinsic attribute of a solution is its electrolytic conductivity, which is a measurement of how well it carries a current – in this case, by ionic carriers rather than electron transmission. The conductance, or G, is a similar parameter that is also utilized and is described as $G = 1/R$. One characteristic of the solution in a particular cell at a certain concentration and temperature is its conductivity. The conductance is dependent upon the cell used to measure the solution; Siemens (S) units are used in G. Three variables affect the electrolyte conductivity: the amounts of ionic species present, their motilities, and ion charges. The following variables have an impact on an ion's mobility or speed in an electrolyte solution:

1. ion radius;
2. viscosity of the solvent (such as water or organic);
3. ion concentration;
4. the applied voltage or strength of the electric field;
5. the solvent's temperature; and
6. the ion's size.

The mobility of the ion is a repeatable physical property under normal circumstances. Since ion concentration is a significant variable in electrolytes, it is common practice to connect electrolytic conductivity to equivalent conductivity:

$$\wedge = K \times V$$

where $\wedge$ is the equivalent conductivity, k is specific conductance, and V is volume in cm^3 containing 1 g equivalent of an electrolyte.

Solutions containing electrolytes behave optimally when one approaches infinite dilution. The reason for this is that as ion concentration rises, so do the electrostatic interactions between ions. The equivalent conductance of the electrolyte Λ nearing infinite dilution is

$$\Lambda_0 = \Lambda_+ + \Lambda_-$$

where Λ_0 is the equivalent conductivity at infinite dilution, Λ_+ is the equivalent conductivity cation, and Λ_- is the equivalent conductivity of anion (table 4 & 5).

According to Kohlrausch's law, the equivalent conductivity of an electrolyte at infinite dilution is equal to the sum of the conductances of its anions and cations. No matter what kind of ion it is associated with, at infinite dilution, dissociation is complete and every ion contributes significantly to the electrolyte's equivalent conductance. Any electrolyte's equivalent conductance at infinite dilution is equal to the sum of the contributions made by each of its constituent ions. The value of identical conductance at infinite dilution for any electrolyte (cations and anions) is stated by Kohlrausch's law.

Table 4: Conductivity of different ions at 25 °C. Petr Vany'sek et al., 2011.

Cation	Λ (Ω^{-1} cm2 mol^{-1})	Anion	Λ (Ω^{-1} cm2 mol^{-1})
H^+	350	OH^-	198
Li^+	38.7	Cl^-	76.3
Na^+	50.1	Br^-	78.4
K^+	73.5	NO_3^-	71.4
Rb^+	76.4	CH_3COO^-	40.9
Cs^+	76.8	ClO_4^-	68
NH_4^-	74	SO_4^{2-}	80
Ag^+	62	$HCOO^-$	55
Cu^{2+}	55		
Zn^{2+}	53		
Fe^{3+}	68		

Analyte	Titrant (NaOH), M	Maximal conductance*
10^{-1} M HAc	1	25 mS
10^{-2} M HCl	10^{-1}	5 mS
10^{-2} M HAc	10^{-1}	2.5 mS
10^{-3} M HCl	10^{-2}	0.5 mS
10^{-3} M HAc	10^{-2}	250 µS
10^{-4} M HCl	10^{-3}	50 µS

*For conductivity cell with a cell constant of about 1 cm^{-1}.

Instrumentation for conductivity measurements

As shown in Figure 11, the apparatus consists essentially of a conductivity cell and Wheatstone bridge. The cell holding the sample makes up resistance A; resistance B is changeable, but resistances D and E are constant. It is possible to modify variable capacitor C and resistor B to achieve zero bridge current or to reach the balance point (see figure 11):

Then,

$$\frac{RA}{RB} = \frac{RD}{RE}$$

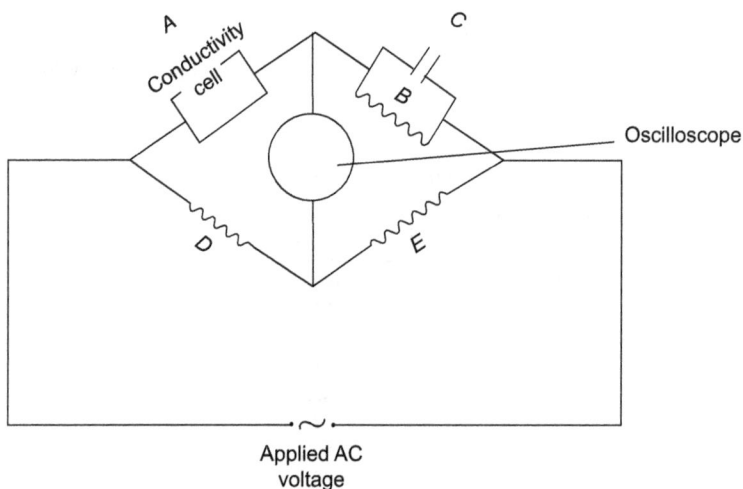

Figure 11: Wheatstone bridge arrangement for conductometric analysis. Malhotra, P. (2023).

Titrations are performed manually point by point, or automatically, where the titrant is introduced continuously (monotonically or dynamically). There are different types of titrations such as acid–base, precipitation, complex formation, and redox titrations. From those, we can now see the application of acid–base conductometric titrations for the experiment. It requires a change in the specific conductance of the investigated solution during the titration in order to find the equivalence point (the amount of titrand and titrant is equal).

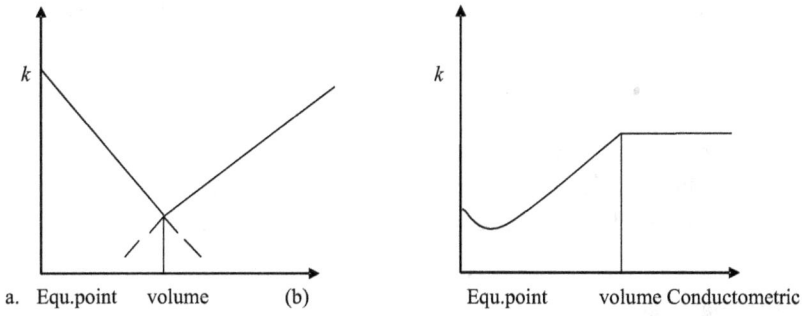

a. Equ.point   volume   (b)   Equ.point   volume Conductometric

Conductometric titration curves for (a) strong acid with strong base and (b) weak acid with weak base.

Experiment 8: conductometric titration of acids and a mixture of acids

Objective
– To get awareness of the technique of conductometric titration.
– To determine the concentration of a strong acid, weak acid, and a mixture of strong and weak acids by means of conductometric titration.

Apparatus and chemicals
Conductivity cell, conductivity meter, magnetic stirrer and bar, 200 mL beaker, burette, 10 mL pipettes, 100 mL measuring cylinder, 0.1 M NaOH, unknown concentration of HCl and CH_3COOH solution, and unknown mixture of HCl and CH_3COOH acids.

Procedure
– Fill the beaker with approximately 90 mL of distilled water after adding 10 mL of the supplied HCl.
– Place the conductivity cell inside and turn on the conductivity meter and stirrer.
– Gradually add 0.5 mL of the 0.1 M NaOH solution to the acid.
– Check the sample's conductivity following each titrant addition.
– Once 10 measurements have been taken above the corresponding mark, continue the titration.
– With the specified mixture and acetic acid of unknown concentration, repeat the same process.

Calculations
– Correct the conductance reading caused by the dilution of the sample during the titration.
– Plot a graph of conductance versus volume of the titrant and determine the equivalent point for all samples.
– Calculate the concentration of both pure acids and the components of the mixture.

Questions
1. After the equivalence point is reached, why is it still essential to titrate?
2. For conductometric tests, why are platinum electrodes that have been plated used?
3. What would happen if direct current was used to test conductivity?
4. What is the reason for conductance to reach minimum at the equivalence point during conductometric titration of a strong acid with a strong base?
5. What are titrant, analyte, and titration?

Experiment 9: determination of chloride in beer samples

Objective
To ascertain the concentration and end point of chloride ions in samples of water and beer.

Apparatus and chemicals
Conductivity meter and dip-type cell with bright platinum electrodes (should be kept clean at all times), beakers (100 mL), pipettes (2 and 20 mL), volumetric flasks, micro-burette with 0.01 mL divisions, $AgNO_3$ (4.79 g/L; (1 mL to 1 mg Cl)) (keep in the dark or wrap aluminum foil around the volumetric flask and standardize by titrating against 2 mL NaCl solution, NaCl (1.648 g/L; (1 mL = 1 mgCl)), and octanol.

Procedure
- To control foaming, degas the sample by pouring it from one beaker to another and adding one or two drops of octanol.
- Pour 20 mL of degassed beer into a 100 mL beaker, top with around 30 mL of distilled water, and mix with the conductivity electrode assembly. Afterward, read and record the conductance measurement.
- Add 0.2 mL of standardized $AgNO_3$ solution, stir well, and record the conductance measurement.
- Add 0.2 mL volume of $AgNO_3$ incrementally until the well-stirred solution's conductivity begins to rise, then note all values.
- Continue adding $AgNO_3$ solution in increments of 0.1 mL until 1 mL has been added after the conductivity starts to rise, and note all values.
- Plot an $AgNO_3$ solution volume as an abscissa on a graph where the conductivity values are represented as ordinates. Two distinct straight lines make up the obtained curve. The titration's end point can be found at the intersection of these two lines.

Calculation of results
1. Standardization of $AgNO_3$, 1 mL $AgNO_3$ = $2/S$ = **P** mg Cl^-, where S is the volume of $AgNO_3$ required.
2. Sample: Chloride concentration, $[Cl^-] = T \times P \times 50$ mg/L, where T is the volume of $AgNO_3$.

3.2 Potentiometry

Potentiometry is one of the most frequently used analytical methods in chemical analysis. In potentiometry, the electric current between two electrodes is kept at almost zero while the voltage difference between them is being measured. The potential of an "indicator electrode" in the most widely used potentiometric methods fluctuates based on the analyte concentration, whereas the potential derived from a second reference electrode (RE) should ideally remain constant. The majority of commonly used potentiometric techniques make use of an ion-selective electrode (ISE) membrane, whose electrical potential to a specific measuring ion – whether in solution or gas phase – produces a highly precise analytical response. There are numerous techniques for determining the end point, including conductometry, amperometry, potentiometry, and spectrophotometry. The most used method is the potentiometric endpoint determination. Using potentiometric techniques, we may determine an analyte's concentration by measuring the potential of an electrochemical cell.

The Nernst equation connects the standard cell potential to the effective concentrations (activities) of the constituents of a cell reaction.

$$E = E° - \ln\frac{RT}{nF}\frac{a_{ox}}{a_{red}}$$

where E is the measured potential, $E°$ is the standard electrode potential, R is the molar gas constant: 8.314 J/mol K at 25 °C, T is the temperature in kelvin, F is the Faraday constant 96,485 Coulombs/mol, ln is the natural logarithm, n is the number of electrons involved in the redox reaction, a_{ox} is the activity of analyte oxidized, and a_{red} is the activity of the analyte reduced.

If the concentration of the analyte is low, which is normal for potentiometric titrations, then the activity $(a) \approx$ concentration of the species. So, the most common form of the Nernst equation is as follows:

$$E = E° - \frac{0.0952}{n}\log\frac{[ox]}{[red]}$$

The application of these relationships is the basis of potentiometric methods. Since the potential of a single electrode cannot be measured, an indicator electrode is always combined with a suitable RE to form electrochemical cells. Under equilibrium conditions, the cell voltage is the difference of the indicator electrode potential and the RE potential. Furthermore, a liquid junction potential (a potential difference that arise at a liquid boundary as a result of different ionic mobilities) may occur. To avoid such a junction, the two electrodes are connected by an electrolyte bridge filled with highly concentrated KCl and NH_4NO_3 solution of a salt whose ions have nearly equal mobilities. Moreover, a liquid junction potential – a potential difference brought about by varying ionic mobilities at a liquid boundary – may manifest itself. The two electrodes are joined by an electrolyte bridge that is filled with a highly concentrated KCl solution and an NH_4NO_3 solution of a salt whose ions have almost similar mobilities in order to prevent such a junction.

Since at a given temperature the potential of the RE is constant, the cell voltage is then

$$E_{cell} = E_{ind} + Constant$$

E_{ind} is the potential of an indicator electrode.

Indicator electrode of the analytical redox titration reaction of

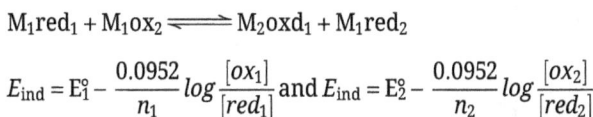

$$M_1red_1 + M_1ox_2 \rightleftharpoons M_2oxd_1 + M_1red_2$$

$$E_{ind} = E_1^o - \frac{0.0952}{n_1} log \frac{[ox_1]}{[red_1]} \text{ and } E_{ind} = E_2^o - \frac{0.0952}{n_2} log \frac{[ox_2]}{[red_2]}$$

At equivalence point, the potential of an indicator electrode is given by

$$E_{ind} = \frac{n_1 E_1^o + n_2 E_2}{n_1 + n'_2}$$

The potentiometric method is ideal for monitoring oxidation reduction (redox) reactions. In such a system, the best indicator electrodes are inert metal electrodes (Pt or Au electrodes) or graphite electrodes and calomel or silver–silver chloride electrodes as REs.

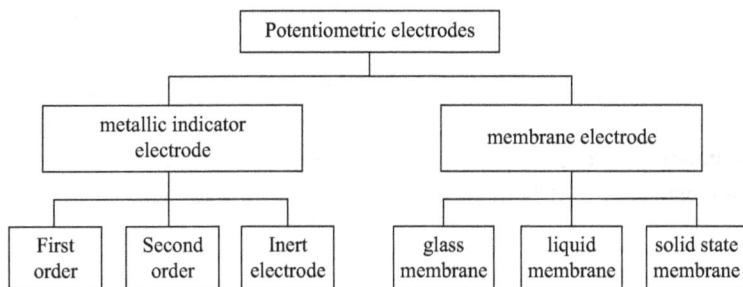

Potentiometric electrodes

- metallic indicator electrode
  - First order
  - Second order
  - Inert electrode
- membrane electrode
  - glass membrane
  - liquid membrane
  - solid state membrane

Different types of electrodes include:
Glass electrodes are combined glass REs with indicator and REs in the same body.

ISEs are used for the detection of specific ions in a mixture of ions. The sensor element, ion-selective membrane, has a construction similar to that of a glass electrode (see figure 12).

Silver indicating electrodes are Ag wires with 1–2 mm diameter. The silver salt precipitate should periodically be removed from the electrode surface when using precipitation titration (either chemically by immersing the electrode in NH_3 solution or manually using fine-grade emery paper). On the other hand, adding a surfactant, such

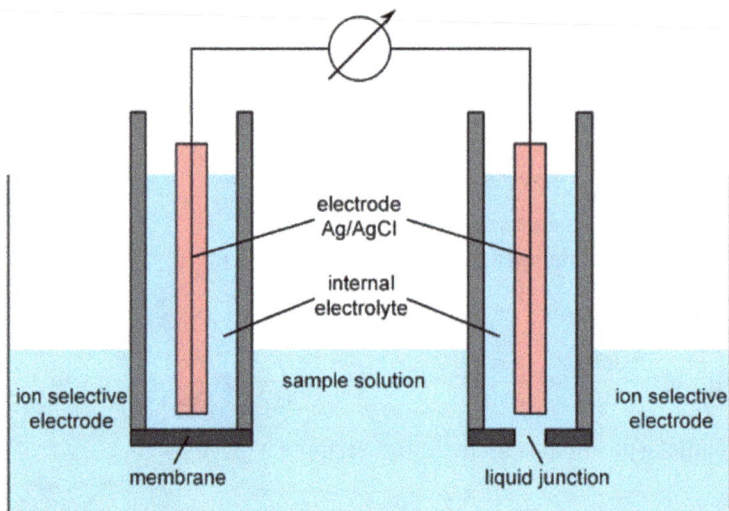

Figure 12: Schematic of ion-selective electrodes.

as polyvinyl alcohol (1 drop of 0.3% PVA to every 5 mL of solution), is a simpler way to stop the electrodes from coating. One method of preparing mercury-coated indicating electrodes is reported to amalgamate a gold wire lightly.

Gold redox electrodes are rarely employed in redox potentiometric titrations; instead, platinum redox electrodes are used. In potentiometric titration, REs such as calomel and silver/silver chloride electrodes are frequently utilized. An electrode made of mercurous sulfate may be utilized in the event where chloride interferences are likely (as in the case of halide determination). The RE for the Ag/AgCl/1 M KCl series of tests that follow is a homemade one. At 25 °C, its potential is −19 mV versus SCE.

Experiment 10: potentiometric redox titration

The process of titrimetric analysis involves calculating the moles of reagent (titrant) needed to quantitatively react with the material under investigation. The titrant can be added in two ways: (a) volumetrically using a low flow-rate pump, glass, or automatic burette; or (b) coulometrically using electrochemical generation from an appropriate electrolyte. Titrimetric analytical techniques have the advantage of being absolute, similar to gravimetric techniques, in that the concentration of the material under investigation is ascertained using fundamental chemical concepts, negating the need for calibration curves (figure 13).

Objective
Using a standard potassium dichromate, K_2CrO_4 solution, titration is used to measure the concentration of an iron(II) solution.

Apparatus and chemicals
Pt electrode, saturated calomel electrode, magnetic stirrer and bar, pH meter, beaker, burette, pipettes, 0.1 N K_2CrO_4 solution, 2.5 M H_2SO_4 solution, and unknown concentration of $Fe(NH_4)_2(SO_4)_2$ solution.

Procedure
–   Fill the beaker with 25 mL of the unknown $Fe(NH_4)_2(SO_4)_2$ solution.
–   Next, add 25 mL of a 2.5 M H_2SO_4 solution and 50 mL of distilled water.
–   Place the 0.1 N K_2CrO_4 electrodes within the burette.
–   Solutions submerging both the reference and the indicator.
–   Start the stirrer and use the pH meter to measure the potential difference between the electrodes.
–   Measure the cell voltage after adding 3 mL of titrant and stir the mixture.
–   Measure the cell voltage after adding another 1 mL of titrant and stir.
–   Add the titrant in increments of 0.1 mL if you are getting near to the equivalence point, and keep going until the equivalence point has been exceeded by at least 1 mL.
–   If you are close to the equivalence point, add the titrant in portions of 0.1 mL and continue until the equivalence point has been passed by at least 1 mL.

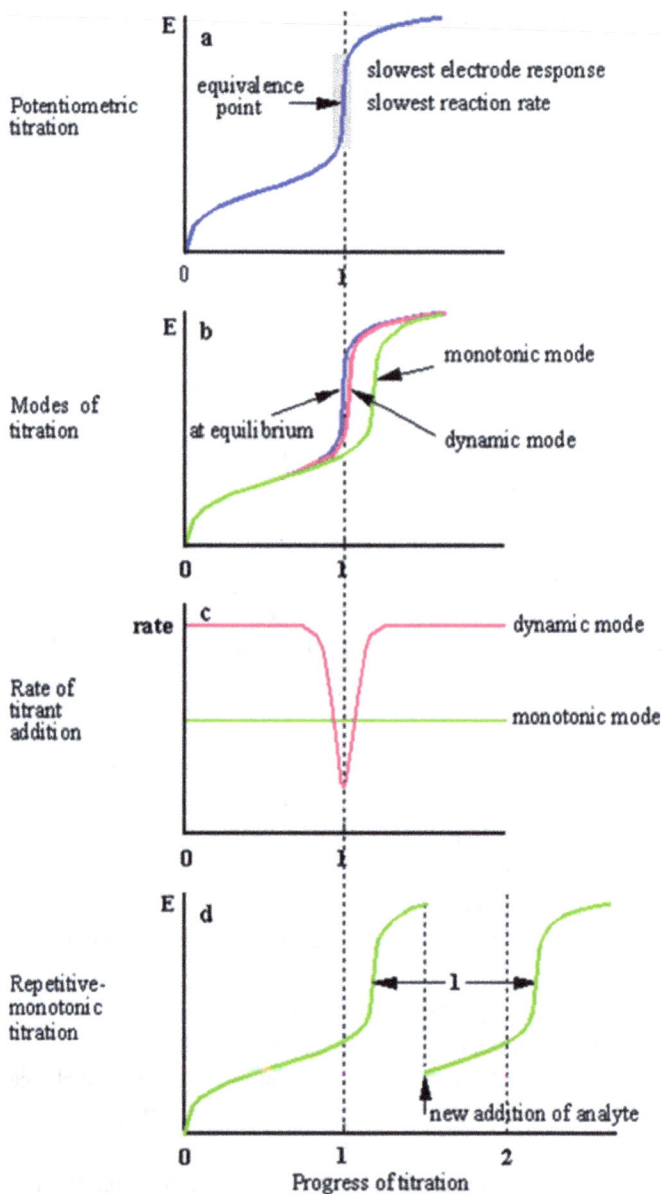

Figure 13: Continuous potentiometric titrations.

Calculation
- Plot the graph of cell voltage versus the volume of titrant.
- Determine the equivalence point (steepest portion of the curve).
- Determine the equivalence point by plotting $\Delta E/\Delta V$ versus V.
- Calculate the concentration of the unknown iron sample and identify the reduced and oxidized forms of species by the reaction of a balanced chemical equation.

Question
1. Write the redox equation for the reaction of Fe^{2+} with $Cr_2O_7^{2-}$.

Experiment 11: potentiometric determination of halides

Objective
– To ascertain from KCl and KI halides, the concentration and K_{sp} of AgCl and AgI halides.
– To demonstrate how the RE affects the halide concentration calculation.

Apparatus and chemicals
Silver wire and mercurous sulfate indicating electrode, beaker, burette, funnel, pipette, 0.05 M KCl, 0.05 M KI, 0.02 M $AgNO_3$, 1 M HNO_3, and 0.02 M $Ba(NO_3)_2$.

Procedure
– It is recommended to clean the silver electrode prior to commencing the experiment. After using fine-grade emery paper to polish, thoroughly rinse with distilled water. Configure the device as indicated below.
– Fill a container with varying amounts of each halide solution (about 3 mL in total), then add 2 mL of 0.02 M $Ba(NO_3)_2$ and a few drops of 1 M HNO_3.
– Take note of the solution's initial volume (±1%); you'll need it to calculate K_{sp}.
– Solve halide titrations using 0.02 M $AgNO_3$.
– Get an unknown sample and perform the titration as described previously.
– At the end of each titration, finally
  1. Plot E versus V (potential versus volume of titrant) and $\Delta E/\Delta V$ versus V.
  2. Report the composition of the unknown in morality.
  3. Calculate the K_{sp} of AgCl and AgI (for a point-by-point titration only; explain the restriction).

Question
What is the purpose of adding $Ba(NO_3)_2$ to the solution?

Experiment 12: potentiometric titration of vegetable sauce

Objective
To estimate the concentration of chloride in a sample of tomato sauce.

Apparatus and chemicals
Plastic beakers, magnetic stirrer, burette, Ag electrode, DC volt, 4M HNO_3, 1.0 g $NaNO_2$, tomato sauce, and Fajan's solution.

Procedure
- Weigh ~2 g (weigh to ±1 mg) of tomato sauce directly into a 100 mL plastic beaker. (Normally you weigh by difference but tomato sauce is difficult to do by this method.)
- Add around 50 mL of distilled water.
- Add about 1.0 g of $NaNO_3$ and three drops of 4 M HNO_3.
- Stir the mixture until the tomato sauce is evenly distributed and the salt has dissolved.
- Lower the stirrer's speed until the solution is gently mixed.
- Dip both the burette tip and the silver electrode into the solution.
- On the multimeter, select the DC volts option and note the potential in millivolts.
- To make the titration process easier, plot the potential as you proceed and note the solution. One person can perform the titration while the other records the results.
- Plot two graphs on a single graph paper sheet: Potential (E) and $\Delta E/\Delta V$ in relation to the volume of additional Ag^+ and Ag^+.
- Calculate the tomato sauce's chloride content and report your findings as % Cl (m/m) and % NaCl (m/m).
- Could you have performed this analysis using Fajan's titration? Fill your 10 mL burette carefully, and position it so that the burette is at an angle. If the solution bubbles down, it will be nearly impossible to remove the bubbles. For your final report which you hand in use a spreadsheet package, for example, *Excel* to calculate $\Delta E/\Delta V$ as well as for plotting your graphs.

3.3 Polarography

Polarography is the branch of voltammetry in which a dropping mercury electrode (DME) is used as the indicator electrode. When electrolyzing solutions containing electrooxidizable and/or electroreducible compounds, polarography is an electroanalytical technique that examines the potential of an electrode and the current that runs through it between a mercury electrode (DME) that is decreasing and an RE. The change in the current flow that results from varying the potential between these electrodes is monitored. The indication electrode is the electrode whose potential is changed. DC polarography at a lowering mercury electrode is the most widely used constant potential method (DME). Voltammetric indicator electrodes can be constructed from a wide range of materials, including graphite, Pt, Au, and mercury, and they can have different constructions and shapes. Plotting the variations in current flow against potential variation yields the following results:

Advantages:
-   The electrode continuously expanded into fresh solution.
-   There is constant renewal of the electrode surface.
-   The reduction of hydrogen proceeds at a high overpotential at mercury.
-   Mercury dissolves most other metals.

Diffusion is primarily responsible for the electroactive species under investigation being transported to the electrode. Other transport modes like migration and convection are suppressed. Migration is suppressed by adding a large excess of supporting electrolytes like 0.1 M KCl. Then, only the supporting electrolyte carries the charges up to the electrode. Reducing external vibrations and operating in an undisturbed solution are two ways to suppress convection. The Ilkovic equation relates the diffusion current to the concentration:

$$I_d = 0.627 \text{ cm}^2/\text{g}^2 \ D^{1/2} nFC$$

where i_d is the average diffusion current n is the number of electrons transferred per a molecule, D is the diffusion coefficient in cm^2/s, rate of flow of mercury in g/s, t is the drop lifetime in second, and C is the concentration of the electroactive species in mol/cm^3 (figure 14).

Polarographic methods
In **direct current polarography (DCP)**, a constant potential is applied during the entire drop lifetime. Depolarization, repolarization, and hyperpolarization are the three primary steps of the action potential. At the conclusion of the drop life, the current is measured. The mercury drop electrode in normal pulse polarography (NPP) is maintained at a constant potential E_{in} over the majority of the experiment, where no electrochemical reaction occurs under the specified experimental circumstances. In the final phase of the drop life, the potential of interest E_p is applied for a considerable amount of time t_p (on the order of several milliseconds) (figure 15). The constant values of E_{in} and t_p are main-

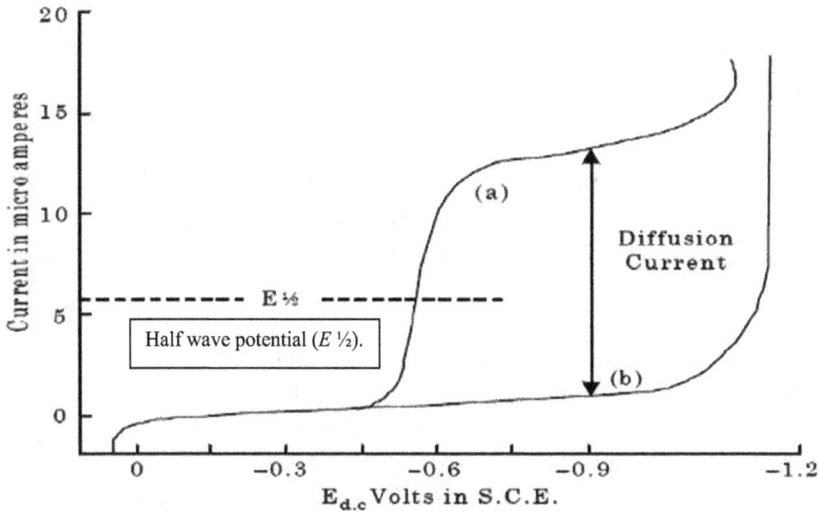

Figure 14: Polarogram: (a) 1 F HCl with 0.5 mM cadmium ion and (b) 1 F HCl alone.

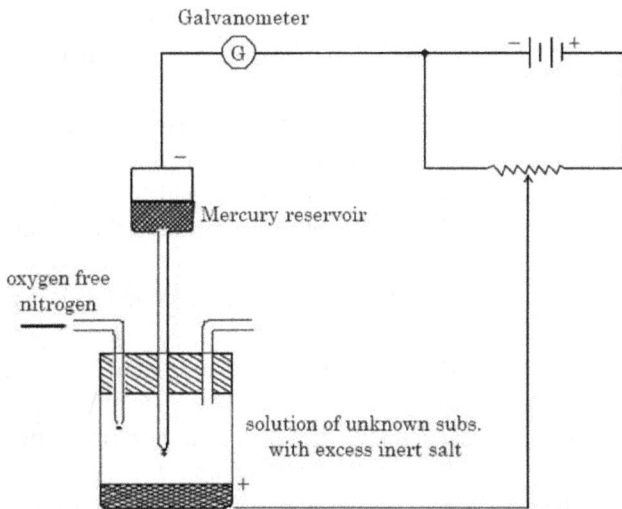

Figure 15: Experimental setup in polarographic experiment.

tained during the E_p, and the polarogram's recording is altered drop by drop. Diffusion control is used in NPP to set the limiting current. Diffusion control has the same experimental criteria as DCP. In contrast, the current for DCP at time τ is

$$I_d = knD^{1/3}m_r^{2/3}t^{1/6}c$$

Here
- With the Faraday constant F evaluated at 708 for maximal current and 607 for average current.
- k is a constant that incorporates π and the density of mercury.
- D is the depolarizer's diffusion coefficient in the medium, expressed in cm^2/s.
- m is the mass flow rate of mercury through the capillary (mg/s).
- n is the number of electrons transferred in the electrode reaction.
- c is the depolarizer concentration in mol/cm^3 and t is the drop lifetime in seconds.

The current under NP conditions at τ is larger than that under DC conditions:
Specifically, the standard polarography experiment has limits when it comes to quantitative analytical measurements. Capacitive current has a significant influence because it is continuously monitored throughout the Hg drop's increase. The surface area initially increases significantly as the mercury flows from the capillary end. Consequently, as charging of the rapidly expanding interface takes place, capacitive effects dominate the initial current. The surface area changes less as the drop life gets closer, which reduces the impact of capacitance variations on the overall current. Any redox reaction that takes place at the same time will produce faradaic current, which decreases roughly as the square root of time (because of the increase). The name "diffusion-limited current" refers to the limiting current (the plateau on the sigmoid) when diffusion is the main factor contributing to the electroactive material's flux at this stage of the Hg drop life. Rather than the waves of traditional polarography, more sophisticated types of polarography (see below) produce peaks, which allow for a greater resolution of various chemical species, and enhance the detection limits, which can in some circumstances be as low as 10^{-9} M.

Polarography using square waves (SWP) yields low detection limits and high sensitivity. DPP stands for differential pulse polarography. From an analytical perspective, DPP's sensitivity is superior to NPP's. The potential sequence utilized to capture a whole differential pulse polarogram, as well as the potential sequence on a single mercury drop, is crucial. The peak's height is directly correlated with the electroactive species' concentration.

Though the complete potential sequence is applied during the lifetime of a single drop, it is comparable to DPP. In comparison to the other methods, which take substantially longer, the voltammogram is obtained in a matter of seconds. For Nernstian processes, the technique performs better than for non-Nernstian processes. The method's detection limit is around 10^{-7} M. The Nernst equation for the DME's equilibrium potential can be used to easily get the formula for the half-wave potential:

$$E = E^\circ + \frac{RT}{nF} \ln \frac{C_{ox}}{C_{red}}$$

where C_{ox} is the concentration of the oxidized form of the species at the electrode surface.

Question: Derive the equation for the half-wave

Any material, solid, liquid, or gaseous, organic compounds with conjugated double or triple bonds, including polynuclear aromatic ring systems, as well as compounds like oximes, imines, ketones, aldehydes, nitrodiazo compounds, and halo-substituted compounds, can be directly determined using polarographic analysis. It is more frequently utilized in disciplines like pharmaceutical chemistry and biochemistry.

Experiment 13: polarography analysis of the polarographic wave

Objective
- To ascertain the half-wave potential and the quantity of electrons participating in the reduction.
- To confirm that the limiting current of the Ti^+ reduction polarographic wave is diffusion regulated.

Apparatus and chemicals
$TiNO_3$, KNO_3, polarogram, stop watch, and volumetric flask.

Procedure
- Prepare 0.005 M solution of $TiNO_3$ in 0.1 M KNO_3. (The Ti^+ concentration has been chosen low enough in order to neglect potential drop in the solution.)
  - To enable an accurate measurement of the currents along the wave, run an additional polarogram at a slow scan rate (5 mV/s).
- Determine $E_{1/2}$ of the Ti^+ reduction wave and compare with the literature.
- Determine the number of electrons involved in the reduction process from E versus $\log [(i_d - i)/i]$ plot.
  - Run a polarogram to choose a potential in the limiting current region. Record a current/time plot at the chosen potential.
- Prove that the current is completely diffusion controlled. Calculate the diffusion coefficient, assuming that $m = 1$ mg/s.

Experiment 14: standard addition method of polarography

Objective
To determine ascorbic acid (vitamin C) in citrus juice using the conventional addition method.

Theory
Ascorbic acid yields a well-defined polarographic oxidation wave. An easily identifiable polarographic oxidation wave is produced by ascorbic acid. Both freshly prepared diluted juice and preserved citrus juice can be used for the determination. Also ascorbic acid (vitamin C) was determined by an electrochemical method from orange juice sample.

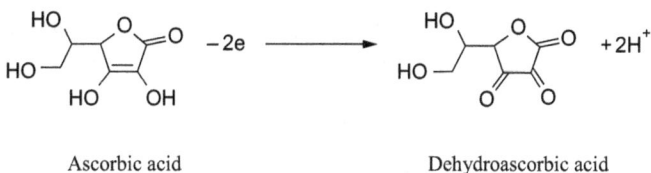

Ascorbic acid   Dehydroascorbic acid

Apparatus and chemicals
Standard solutions of ascorbic acid, 0.5 M acetate buffer, orange, grape fruit, lemon, 25 mL volumetric flasks, porous funnel, and polarography.

Procedure

Calibration curve
- Prepare a fresh stock solution of 50 mL 0.2% ascorbic acid.
- Prepare five standard solutions of ascorbic acid in volumetric flasks of 25 mL. Add 0.5 mL 0.5 M acetate buffer and different volumes of 0.2% ascorbic acid: 0, 200, 400, 600, and 800 µL.
- Dilute to the mark with distilled water to each one.
- For each solution record a NP ($\tau = 1s$, t_p = 20 ms, pulse amplitude = 20 mV, scan rate = 10 mV/s) and SW ($\tau = 2s$, t_p = 20 ms, pulse amplitude = 20 mV, scan rate = 100 mV/s) polarograms over the potential range: −150 to +200 mV versus Ag/AgCl/ 1 M KCl, with E_{in} = −150 mV.

   Consult with the instructor about the preferable mode of polarography to be used in further experiments.

– Plot i_d versus concentration of ascorbic acid. Is the plot linear and does it pass through the origin?

On the basis of these observations, decide if the standard addition method is applicable.

Determination of ascorbic acid in citrus fruits

Squeeze the juice out of an orange, grapefruit, or lemon until around 10 mL is produced. Pour the juice through a funnel with porous material (approximately 1 mm in diameter). Four 25 mL volumetric flasks should be ready. Pour 2.0 mL of the juice, 0.5 mL of the 0.5 M acetate buffer, and the standard additions of 0, 200, 400, and 600 μL of 0.2% ascorbic acid into each. Adjust the concentration with deionized water. Polarograms should be taken under the same settings as during the calibration phase. Calculate the analyte's concentration by drawing the standard additions plot. Provide the ascorbic acid (vitamin C) concentration in mol/L and ppm for the original sample (juice).

Determination of ascorbic acid in conserved citrus juice

Arrange an experiment to find the ascorbic acid content of preserved (commercial) citrus juice. Utilize the prior experiment's scheme. The removal of water, fruit pulp, and flavors which are re-added in a subsequent production stage following transportation is the primary distinction between pure juice and concentrate.

Experiment 15: polarographic metal analysis

Objective
To illustrate the polarographic technique's analytical viability for metal analysis in waste and natural waters.

Apparatus and chemicals
Three volumetric flasks (100 mL), pipettes (1, 5, 10, and 15 mL), 6 M HNO, 0.0013 M KCl, 0.2% Trito X-100 (surface acting agent), 0.2 M HCl supporting electrolyte, and 0.02 M CdCl are among the equipment used in the recording of polarography.

Procedure
- Make three dilutions of the $CdCl_2$ solution: 0.001, 0.002, and 0.003 M. Pipette precisely 5 mL, 10 mL, and 15 mL parts of the 0.02 M $CdCl_2$ standard solution into each of three distinct 100 mL volumetric flasks. Then, precisely dilute the portions with a supporting electrolyte of 0.2 M HCl.
    - After deaerating each solution, measure the polarogram of each one against the saturated calomel electrode between −0.1 and −1.2 V.
- Find each solution's half-wave potential ($E_{1/2}$), which is the polarography's point of inflection.
- From the polarograms of the 0.01 M, 0.02 M, and 0.003 M $CdCl_2$ solutions, calculate the diffusion current. Plot the diffusion current's value versus the $CdCl_2$ solution concentration.

3.4 Coulometry

A collection of techniques based on the electrolytic oxidation or reduction of an analyte is referred to as coulometry. The process of electrolyte can be used to quantitatively change the analyte's oxidation state or it can be carried out by adjusting the current or voltage. The purpose of controlled potential is to remove interferences caused by reactions occurring at different potentials. Electrogravimetry is a type of coulometry that involves reducing, plating out metallic components onto an electrode, and weighing the results. The use of Faraday's first law of electrolysis, which states that the amount of electricity (current) flowing through an electrode directly correlates with the degree of chemical reaction there, is known as coulometric analysis. It is commonly known that an electroactive analyte needs 1 F (96,485 C) of electricity to reduce (or oxidize) 1 g-equivalent weight of the analyte. The amount of analyte can be determined by measuring the amount of electricity required to completely reduce (or oxidize) a given sample, assuming the reaction is 100% efficient (or of known efficiency).

In original, concentration of the metal in the electrolyte solution is indicated by the weight of the metal material that is formed or deposited. Therefore, the formula gives the weight consumed in an electrolysis employing Q coulombs:

$$W = \frac{M \times Q}{96,487\ n}$$

where M is the relative atomic (or molecular) mass of the substance liberated or consumed, n is the number of electrons transferred in the electrode reaction, W is the mass of the product of electrolysis, and F is the Faraday constant (96,485 C).

The current applied is much larger than the current in polarography, and the duration of a coulometric experiment ranges from minutes to hours. The working electrode must be unpolarized during the reaction so that the electrode reaction is not inhibited. Coulometric methods, named after the coulomb force, are analytical techniques that rely on measuring an electrical quantity and using the aforementioned equation. In order to apply Faraday's law to represent the amount of a substance reacted from the observed quantity of electricity (coulombs) passed, a coulometric analysis requires that the electrode reaction used for the determination proceeds with 100% efficiency. The primary coulometric analysis involves the substance being determined directly undergoing a reaction at one of the electrodes; secondary coulometric analysis involves the substance reacting in solution with another chemical produced by an electrode reaction.

There are two coulometric approaches available: controlled potential and controlled current.

Controlled current method: a steady current is used to electrolyze the substance under investigation. The charge is computed using the equation and the amount of time needed for a full reaction is measured:

$$q = it$$

The potential at the working electrode changes (a working cathode becomes steadily more negative, and a working anode more positive) as a result of the sample's change in composition during the coulometric experiment, and a side reaction may occur. The potential must not be greater than the value at which an electrode reaction occurs in another species in order to prevent this reaction. The so-called secondary constant current coulometry or coulometric titration – a technique that produces a reagent or titrant that reacts quickly and stoichiometrically with the analytical species – can be used to get around this.

A solution of the substance to be determined is electrolyzed with constant current in the constant current coulometry method until the reaction is finished (as indicated by a visual indicator in the solution or by amperometric, biamperometric, potentiometric, or spectrophotometric methods), at which point the circuit is opened. The formula for calculating the total amount of electricity passed is current (amperes) × time (seconds). Currently, an electronic integrator is incorporated into the circuit.

Managed potential techniques: The working electrode's potential is set to a value where only the reaction under investigation occurs quickly. The material under investigation reacts at a working electrode with 100% current efficiency. The reaction's completion is indicated by the current dropping to almost zero over time. The amount of the material reacted is measured using a current–time integrating device or a coulometer connected in series with the cell. In a controlled-potential coulometric analysis, the current generally decreases exponentially with time according to the equation

$$I_t = I_o e^{-k't} \text{ or } I_t = I_o 10^{-kt}$$

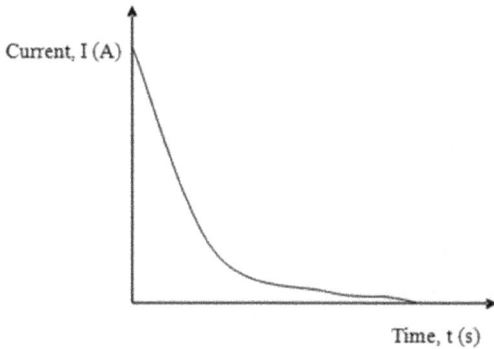

The charge used for the electrolysis from the beginning of the determination to time is given by the equation

$$Q = \int_0^t I(t)dt$$

where **I** is the current at time **t**.

The above equations relating the variation of current with time can also be expressed in terms of the concentration of electrolyte at time t (C_t), and the initial concentration (C_o), that is, $C_o e^{-kt} = C_t$. Since this equation represents a first-order reaction process, the initial concentration has no bearing on the fraction of material electrolyzed at any given time.

The integration can be carried out automatically by a mechanical current–time integrator, or graphically. The electrolysis process in regulated current coulometry is limited to a very small constant value (≤0.01 A), but not to zero. In an experiment where only the supporting electrolyte is electrolyzed, this background current (I_b) is measured. Next, the mass of the sample is computed using the following equation:

$$m = \frac{M(Q - I_b t))}{nF}$$

The primary advantage of coulometry is its high accuracy, which can be accurately controlled and monitored because the "reagent" used is electrical current. Commercial equipment use coulometry to generate both unstable and stable titrants "on demand," study redox reactions, and evaluate fundamental constants in addition to continuous analysis and process management of the manufacture of chlorinated hydrocarbons. Additionally, it is employed to continuously check the amount of mercaptan present in the raw materials needed to make rubber. Another interesting use of coulometry is the production of a chemical reagent in solution. Acids, bases, oxidizing and reducing agents, and other reagents used in volumetric analysis can all be made in solution and permitted to undergo reactions.

Constant current coulometry apparatus
The continuous current setup. The components of coulometry include an electrolysis cell, an end point detector, a current measuring device, and a constant current DC power supply.

Table 5: Mercury cathode's regulated potential deposition of a few metals.

Element	Supporting electrolyte	Volts versus S.C.E.	
		$E_{1/2}$	$E_{cathode}$
Cu	0.5 M acid sodium tartrate, pH 4.5	−0.16	−0.09
Bi	0.5 M acid sodium tartrate, pH 4.5	−0.23	−0.40
Pb	0.5 M acid sodium tartrate, pH 4.5	−0.48	−0.56
Cd	1 M NH_4Cl + 1 M aq. NH	−0.81	−0.85
Zn	1 M NH_4Cl + 1 M aq. NH	−1.33	−1.45
Ni	1 M pyridine + HCl, pH 7.0	−0.78	−0.95
Co	1 M pyridine + HCl, pH 7.0	−1.06	−1.20

General technique: A coulometric determination at controlled potential of the mercury cathode is performed by the following general method.

The general approach that follows is used to produce a coulometric determination of the mercury cathode at a regulated voltage. After adding the supporting electrolyte (50–60 mL) to the cell, air is forced out of the mixture by rapidly circulating nitrogen through it for roughly 5 min. The mercury reservoir is then raised, allowing the cathode mercury to be introduced through the stopcock located at the bottom of the cell. After starting the stirrer, the RE's bridge tip is adjusted so that it barely contacts or trails into the churned mercury cathode. Nitrogen is continually passed through the solution during electrolysis, and the potentiostat is adjusted to maintain the correct control potential until the current drops to a very small constant value.

After roughly 10 min, the current typically drops to 1 mA or less as a result of this initial electrolysis, which eliminates any remnants of reducible contaminants. After pipetting a known volume of the sample solution – roughly 10–40 mL – into the cell, the electrolysis is allowed to continue until the current falls to the same minimal amount as was observed when the supporting electrolyte was used alone. Usually, electrolysis takes an hour to finish. Using the relevant formulas above, the electronic integrator is finally read and the weight of metal deposited is computed.

Experiment 16: separation of nickel and cobalt by coulometric analysis at controlled potential

Objective
To separate Ni and Co metals from a given electrolyte solution.

Apparatus and chemicals
Purified ammonium nickel sulfate (10 mg/mL) and pure ammonium cobalt sulfate (10 mg/mL) standard solutions, and pyridine (redistill pyridine and collect the middle fraction boiling within a 2 °C range, i.e., 113–115 °C). A small background current is obtained with the latter supporting electrolyte, which is composed of 1 M pyridine and 0.5 M chloride ion, adjusted to a pH of 7.0 ± 0.2 for use with a silver anode, or 1 M pyridine, 0.3 M chloride ion, and 0.2 M hydrazinium sulfate, adjusted to a pH of 7.0 ± 0.2 for use with a platinum cathode.

Procedure
- Take 90 mL of the supporting electrolyte in the cell, replace the dissolved air with pure nitrogen, and apply first electrolysis to the solution using the mercury cathode's potential.
- To eliminate residues of reducible contaminants, use 1.20 V versus S.C.E.; once the background current (2 mA) has dropped to a constant value, 30–60 min should pass.
- Ten milliliters of each of the prepared solutions should be added to the cell along with the coulometer and the potentiostat adjusted to keep the cathode potential at the value to be utilized in the determination (−0.95 V vs S.C.E. for nickel).
- Electrolyze until the current drops to the background current value, then record how many coulombs are passed.
- Determine the weight of the deposited nickel (Ni).
- Determine how much cobalt (Co) is there.
- If required, each determination can have its background current adjusted by deducting the amount $I_b \times t$ from Q (number of coulombs recorded), where t is the electrolysis duration in seconds and I_b is the base current.

Experiment 17: colorimetric titration method

Objective
To determine the concentration of chlorine in beer and water sample by coulometric titration.

Apparatus and chemicals
Conical flask (250 mL), pipettes (10 mL, 25 mL), burette, indicator paper, and indicator (0–5% diphenylcarbazone in ethanol, 1% w/v diluted nitric acid, and mercuric nitrate solution). The chloride standard is 1.648 g NaCl dissolved in water to form 1 L (1 mL contains 1.0 mg chloride ion). Dissolve 3.1 g mercuric oxide in the least amount of diluted nitric acid and make up to 1 L with distilled water.

Procedure

Standardization
A 250 mL conical flask should contain 10 mL of the chloride standard, 40 mL of distilled water, and 0.5 mL of indicator. Add diluted nitric acid drop by drop until the indicator paper's pH equals 2.5. Titrate to a permanent (slight but noticeable) purple color using mercuric nitrate. S mL was utilized, and $10/S$ mg of chloride ions is equal to 1 mL of mercuric nitrate.

- Pipette 25 mL wort or degassed beer into a 250 mL conical flask and add about 25 mL distilled water in one of the first conical flask, and in the second conical flask, add 50 mL of water samples and combine with 0.5 mL of indicator.
- Titrate with mercuric nitrate (say T mL, the volume of mercuric nitrate solution consumed for both the beer and water samples) after adding dropwise diluted nitric acid till pH = 2.5 (indicator paper).
- **Question:** What is the chloride content?

References/texts

[1] G.H. Jeffery, J. Bassett, J. Mandham and R.C. Denney, Vogel's Text Book of Quantitative Chemical Analysis, 6th ed., John Willey and Sons, Inc., New York 2000.

[2] D.A. Skoog and J.J. Leary, Principle of Instrumental Analysis, 4th ed. Saunders College Publishing, 1992.

[3] G.D. Christian, Analytical Chemistry, 5th ed., John Willey and sons, Inc., New York, 1994.

[4] F.W. Fifield and D. Keale, Principles and Practice of Analytical Chemistry, 3rd ed., Blakie academic and professional, Glasgow, 1990.

[5] J.M. Marmet, M. Otto and H.M. Widmer (editors), Analytical Chemistry, Willey-VCH, Weinheim, 1998.

[6] chemistry/groups/Li/chem316/Chem316labmanual_2003-3.pdf

[7] https://www.tau.ac.il/~advanal/Polarography.htm

[8] Malhotra, P. Instrumentation for conductivity (2023).

[9] Petr Vany'sek et al., conductivity 2011.

[10] Sambrook, J., Fritsch, E.F., and Maniatis T, sodium dodecyl sulfate–polyacrylamide gel electrophoresis (1989).

[11] Marek Minárik, Analytical Science (Third Edition), 2019.

[12] John V. Hinshaw, thermal conductivity, 2006.

[13] Byjus et al., Paper chromatography, 2009.

https://doi.org/10.1515/9783111575636-005

Appendix

Recipes for selected acid–base indicator solutions

Methy1 violet	0.01–0.05% in water
Cresol red	0.1 g in 26.2 mL of 0.01 M NaOH + 223.8 mL of water
Thymol blue	0.1 g in 21.5 mL of 0.01 M NaOH + 228.5 mL of water
Methyl orange	0.1% in water
Bromocresol green	0.1 g in 14.3 mL of 0.01 M NaOH + 235.7 mL of water
Methyl red	0.02 g in 100 mL of 60% v/v ethanol–water
Bromothymol blue	0.1 g in 16 mL of 0.01 M NaOH + 234 mL of water
Phenolphthalein	0.5 g in 100 mL of 50% v/v ethanol–water
Thymolphthalein	0.04 g in 100 mL of 50% v/v ethanol–water
Clayton yellow	0.1% in water

Source: Reprinted from CRC Handbook of Chemistry and Physics, 82nd ed., Copyright CRC Press, Inc., Boca Raton, FL, 2001–2002. With permission.

Concentration data for commercial concentrated acids and base

Acid or base	Molarity	Density (%)	Composition (w/w)
Acetic acid ($HC_2H_3O_2$)	17	1.05	99.5
Ammonium hydroxide (NH_4OH)	15	0.90	58
Hydrobromic acid (HBr)	9	1.52	48
Hydrochloric acid (HCl)	12	1.18	36
Hydrofluoric acid (HF)	26	1.14	45

If molarity is to be calculated:

$$L_T X \; M_T X \text{ mole ratio } (PS/T) = \frac{\text{grams}_{PS}}{WF_{PS}}$$

If normality is to be calculated:

$$L_T \times \; N_T = \frac{\text{grams}_{PS}}{EW_{PS}}$$

If the molarity of the titrant is to be used:

$$\% \text{ anayte} = \frac{L_T \times M_T \times \text{mole ratio}(ST/T) \times FW_{\text{analyte}}}{\text{Weigh to fsample}} \times 100$$

https://doi.org/10.1515/9783111575636-006

If the normality of the titrant is to be used:

$$\% \text{ analyte} = \frac{L_T \times N_T EW_{analyte}}{\text{Weight of sample}} \times 100$$

Back titration, if normalities are used:

$$\% \text{ analyte} = \frac{(L_T \times N_T - L_{BT} \times N_{BT}) \times EW_{analyte}}{\text{Weight of sample}} \times 100$$

Index

https://doi.org/10.1515/9783111575636-007

www.ingramcontent.com/pod-product-compliance
Lightning Source LLC
Chambersburg PA
CBHW081551220326
41598CB00036B/6640